YAHWEH'S DIVINE

WAR CHARIOTS

By

C. PRESTON BOST

ISBN: 1-4033-8820-2 (e-book)
ISBN: 1-4033-8821-0 (Paperback)

Library of Congress Control Number:
2002095311

This book is printed on acid free paper.

Printed in the United States of America
Bloomington, IN

1stBooks – rev. 03/06/03

Many of the Bible references used have been paraphrased for clarity and as an aid to understanding. *Yahweh*, through his inspired writers, had his word recorded in the manner of spiritual discernment. Anyone lacking true knowledge will never be capable of achieving an accurate interpretation of his divine revelations to mankind.

CONTENTS

INTRODUCTION

Without question, the most monumental miscalculation the civilized world of mankind has ever made is his continued arrogant rejection of the spirit alien visitation reality (an undeniable reality that has been witnessed by literally millions of people around the world from all walks of life and professions).

In convincing evidence, the historical biblical record clearly shows that alien spirit beings have been visiting our planet earth for thousands of years now. Indisputable evidence also provides detailed, descriptive information on extra-terrestrial spiritual beings, as well as the spacecraft they use for inter-galactic travel among the billions of constellations throughout the magnificent universe.

Most uninformed people of the earth have been blindly misled into believing that spiritual beings (angels), are nothing more than mythical beings or entities which have very little, if anything, to do with their own lives, and even less influence over mans' future destiny. Not so: *Yahweh*, the grand creator and "first cause" of all that "exists" has long ago commissioned his angelic beings to fulfill his divine will and purpose in both the celestial heavens and the planet earth. This commissioned fulfillment is taking place in and during this present "rejected" generation of mankind.

These spiritual warriors of the only true God, *Yahweh,* will soon awaken all of mankind from a long slumber of rebellion and denial.

For the last thirty-five years, I have conducted an in-depth research study into the UFO/Alien visitation reality, and how the UFO phenomena relates to the spiritual realm, and its potential future influence over the "modern civilized world" of mankind and his final destiny.

The following synopsis will present my summary and unbiased conclusion of my research findings. I regretfully acknowledge that many will take exception to the conclusions expressed. However, I have made every effort to be straightforward in my interpretation of the truths and facts as investigated. Be informed: We are not alone in this grand universe, or here on earth. "Alien" spirit beings are real, and they are regularly visiting us. These angelic "creatures" are being sent here to carry out the original pronouncements of *Yahweh*. This is clearly stated in his word at Genesis, Chapter 2. His angels are now conducting preliminary preparation work for the restoration of the earth to its original garden-like state. Their worldwide cleansing of this planet will soon affect every living thing on this earth. The Bible record unequivocally reveals *Yahweh's* intended purpose for mankind's final state of existence.

Yahweh sent his highest angelic being, Christ Jesus (Michael), to the earth from the spirit realm to reclaim mankind back into the universal community. As long as mankind carries the stain of sin, he has no hope for continued life on earth.

Some may reason that the act of materializing into human flesh from the spirit realm is an impossibility. This transformation is well within the biological capability of *Yahweh*, who created both spirit and flesh. The spirit realm is by far the more important creation, as his angels were used in the creation of the "physical world." I therefore challenge all to make a diligent examination of the recorded facts and live a life of eternity.

Yahweh's message to mankind, contained in his written word, was provided for man's own good. Man's understanding of his written word today is noticeably lacking, even among those who profess to be "true Christians." This lack of Bible knowledge and understanding leaves very little foundation to anchor their convictions or hope for the future upon. One's ability to resolve everyday problems of life, and in being able to make proper decisions in times of crisis, has become more and more difficult. Most people have allowed themselves to become so weighted down with anxieties of life, that they have little time to study the Bible. Still, the word of God is the most absorbing, interesting, and profitable book we could ever read or study on earth. A sincere, in-depth search of his revelation to man can open the way to true knowledge. This will help us to appreciate the full power of the spirit creation, and its direct connection between the spirit realm and the world of mankind.

Basically, knowledge is the familiarity with reality and the facts, which is acquired through personal experience, study, or observation. Accurate knowledge of the divine word is more valuable than choice gold. Can you imagine the wisdom we would possess if we would reach the perfect knowledge of God?

The role that *Yahweh* assigned to his spirit angelic son in fulfilling his will and purposes is of such extraordinary importance that *Yahweh* concealed in his first spirit son, "Michael" (Jesus), "All the treasures of wisdom and of knowledge." Unless we exercise "faith" in Jesus Christ as God's only begotten son, we cannot come to know "divine wisdom."

Frequently, Bible knowledge is associated with other desired Christian attributes, which include:

"Wisdom."
Wisdom is one's ability to use knowledge intelligently, or the effective application of what is learned. The person may have considerable knowledge, but does not know how to apply it because they lack wisdom.

Jesus said, "Wisdom is proved righteous by its works." Without wisdom, knowledge is of little value.

"Understanding."
Understanding is the ability to see how minor parts or aspects of a matter relate to each other, and the ability to envision an entire matter, not just isolated facts. The basic mental ability to separate, or distinguish, or to discern points or thoughts related to a subject, and fitting them together in logical meaning or order, and ability to join new information to what has already been learned.

"Discernment."
Discernment is the seeing of, and recognition of things as they really are, or comprehending truth and reality, no matter the consequences. Basically, it is weighing or evaluating things in light of truth or reality of a matter.

"Thinking Ability."
Thinking ability is closely related to knowledge. It is best described as the ability to control the mind in directing our thinking and

thoughts in proper, admirable, upright channels, thereby directing our thoughts in agreement and in harmony with godly wisdom and knowledge.

ACKNOWLEDGMENTS

I would like to express my sincere appreciation to all those who so generously shared their extra-terrestrial confrontational experience with a divinely directed celestial space vehicle. Your contribution of detailed information relating to your remarkable, but rare encounter is an experience few people are granted, and even fewer still, understand. The data you provided relating to the unexampled space vehicle phenomena formed the very basis for this book, and opens an extremely advanced scientific field of study and research.

FOREWORD

GOD'S WAR CHARIOTS

This is perhaps the first book ever written to seriously explain the UFO origin and the alien visitation reality. Many have speculated about the UFO phenomenon, but all have failed in every instance to explain their origin factually. Such intentional denial of the alien reality has confused and failed to convince the world of mankind who these "strange earth visitors" really are.

The most commonly accepted reaction today is to repudiate, ignore, and deceive the general public as to the reality of this supernatural alien existence. An abundance of indisputable evidence clearly establishes their presence in the earthly realm.

The entire writing of this book has been devoted to an unbiased evaluation and explanation of the visiting aliens' hidden mission in coming here, and why this can be "the only truthful reality". It explains why mankind is the missing link in the universal chain of celestial brotherhood, and how man, in his inordinate lust for fame and social power, is never satisfied, even when he has gained worldwide notoriety, fame, and incalculable wealth. He is never willing to compromise his fanatical "power position" in society, regardless of the consequences. It is this overmastering desire for wealth, power, control, authority, and independence that has driven human society to the brink of impending destruction. Ironically, the preparatory purification process of human society being carried out by the aliens (spirit beings) must be completed before our newly transformed planet, earth, can be admitted into the universal association of men and angels.

ABOUT THE COVER

This is an artist depiction, painted by Charles Preston Bost, of an unusual, unexplained spacecraft sighted near Statesville, North Carolina in five separate encounters by six individual creditable witnesses. Observer testimony is in full accord with details of the craft as depicted by the painting. All the observers described the large silent craft as an "unearthly" apparition.

The painting resides in the Roswell Science and Research Museum in Roswell, New Mexico.

CHAPTER 1

A DIVINE RECORD

Yahweh, sovereign ruler over his creative works, in his inspired written word, revealed to mankind many thousands of years ago that he had made living beings both in the heavens and on our planet earth.

Simple observation clearly proves his creative process to have been a progressive work of unsurpassed wisdom, power, and technological wonder. In fact, his creative works are so vast in scope and magnitude that man is unable to even comprehend the extreme size and complexity of our own "small" galaxy, let alone the billions of other galaxies, which make up the universe itself.

Ultimate wisdom can be seen in the delicate but precise balance installed in the universal laws. If we were to examine the four universal forces of the creation, we would discover that these four basic forces keep the various elements, which make up billions and billions of heavenly bodies, in perfect balance and

1

consistency. The four universal forces are gravity, electromagnetism, and the strong and weak nuclear forces.

The finely tuned balance, consistency, and dependability of these four forces make all life possible on earth and in the unlimited heavens. By divine wisdom, foresight, and intelligence, *Yahweh*, the "One" who represents technology personified, also made earthling man and spirit angelic beings. All living beings, both fleshly and spiritual were created to inhabit separate environmental realms. Both were made to exist to fulfill *Yahweh's* commission to care for and watch over his "works."

Significantly, *Yahweh*, through his inspired prophets on earth informed us that; "There were heavenly bodies and earthly bodies." He also tells us, "The glory of the heavenly body is of one sort (angels), while the earthly bodies are of a different sort (flesh)."

This revelation to man is very revealing and can be directly associated with spirit being visitation to earth (UFO enigma). By making this connection, we have the technical expertise to make alien being visitation to planet earth a practical reality. Since spirit beings were created billions of years prior to man, is it any wonder these heavenly creatures (angels) would have the advanced technical capability to transcend time, distance, and space?

Why are the angelic beings coming to earth? Where do they come from? Are they "being sent" here on a mission? These are but a few questions raised by the reality of alien being visitation. Such far-reaching questions should be of sincere interest and concern to every living human being living on this beautiful earthly home of mankind. Why? We should all be concerned, because the future existence of man is under full control of *Yahweh's* heavenly beings. They have been commissioned and given "all authority" to carry out the "First One's" will and purpose. His purpose in placing man on this planet was not for man to make war against our fellow man. He instructed us in what he expected of us, and has not, and will not change his will and purpose for mankind. Man has yet to accept this fact. As a result, man has so far failed to gain purposeful discernment of his own existence because of his pride and rebellious attitude.

It is clear that the scientific community has deliberately ignored the fact that reliable witnesses report many thousands of unidentified spacecraft sightings each year. Many reported occupant observations in or near the craft seen, yet the scientific world continues to deny this is even happening. They will not accept the reality that aliens have the ability and authority to materialize, or transform themselves into *any form* necessary to fulfill their mission as directed by *Yahweh*.

This chosen blindness and lack of true understanding within the scientific community is masking the *real truth* from the "ordinary" people who are content in most instances with the "lie" of deception. This fatal failure by mankind will ultimately result in the total destruction of this world's system of rebellion by choice.

Continued covert denial of indisputable evidence of the alien being reality does not nullify the truth of visitation. Scientists and the so called other "learned professionals" will no doubt continue to explain away the reality no matter how provable the evidence presented, or how convincing the facts may be. This delusion is not founded upon a lack of hard evidence, but is rejected solely because the scientific establishment recognizes that the alien phenomena are well beyond the present scientific understanding. If you will, please read the Bible Book of Job. This will clearly show you where the world of *high science* stands in relation to true knowledge and understanding.

Is it scientific fact to say that something does not exist because it cannot be explained? And yet science leaders are in effect saying this, even though the evidence is seen by thousands of people all over the earth. This is the unreasonable position the scientific establishment has taken with regard to extra-terrestrial life and earth visitation. Most scientific critics say they will believe extra-terrestrials exist only when an alien says, "Take me to your leader." Little do they know that the leader, *Yahweh*, is sending the aliens to earth. "World leaders" do not understand the true significance of extra-terrestrial visits. They will not admit to the superiority of the angelic intellect. Scientists and political leaders do not understand a simple fact: The spirit beings now visiting earth have no interest in "making contact" with world leaders. They have their "marching" orders from *Yahweh*, and are here to fulfill their mission in the appointed time designated. They deny the alien presence because they do know that these unexplained phenomena can have a worldwide impact upon mankind's system of things. This will directly affect rulership and class manipulation. King David gave us an insight into the magnitude of the heavenly angelic armies. He recorded; "The war chariots of *Yahweh* are tens of thousands, and thousands and thousands, over and over again." That statement represents millions of war chariots (UFOs?). Does it not? These faithful revelations about his heavenly forces are absolutely mind-boggling, but are in fact reality; A reality no one can *intelligently* deny. A "war chariot" is in fact, a vehicle used by warriors in battle, also known as a "war wagon."

Yahweh, a divine manly leader of warfare, and the "God" of heavenly armies, has every right, as creator and supreme ruler over the universe, to execute justice against all who refuse to recognize Him as the highest sovereign God and Judge. The Great War of God the Almighty, soon to be fought here on the earth involving the angelic warriors, will not be a local skirmish. This war will be global in scope and will include *all* the kings of the earth *and* their armies of the "entire inhabited earth." They will be led by the inspired expressions of angelic demons (Revelations 16:14). The true followers of *Yahweh* will have no part in this warfare, but will be mere onlookers of this final battle ever to be fought on planet earth.

CHAPTER 2

THE FIRST CAUSE

Anyone who is willing to make a diligent, sincere search of the Biblical historical record, will, without a doubt, conclude that the visible, physical realm of which mankind is a part of, was in fact, brought into existence thousands of millions of years after the creation of spirit beings by "*Yahweh*", the First Cause. Because of this enormous gap in time, it is far beyond man's thinking capability to comprehend what the distant past may have been like, or what the distant future may hold. Man can only *know* what the *present* age of his own civilization has attained. To *know* beyond that, he must rely on the recorded historical record, or the archaeological fossilized formations. It is extremely difficult for mankind to understand that *time* has neither beginning nor end. He can only comprehend "real" time as that: Time he experiences in his own short lifetime (life span). He is particularly aware of time because he *knows* his years are but *few* relative to *eternal time*. Time as measured from any given point in time is eternal as measured back in the past or as measured in the future time. *Yahweh* gave mankind a very revealing bit of information about time when he said, "I am from time indefinite, to time indefinite." In other words, he was letting us know that he had no beginning and will have no end. He has been "from forever to forever." We as human beings are subject to "death" only because our "forever" rests in "the forever." Death is a suspension of our "forever." In order for mankind to regain his "forever" (eternal life) he must first regain *Yahweh's* favor. Our "forever" can never be restored by scientific breakthroughs. Only through the mercy and undeserved kindness of "The First Cause" (*Yahweh*), can we receive the hope of life eternal.

By supreme wisdom, The First Cause brought all things into existence with a predetermined purpose. He then commissioned his created beings, both spirit and flesh to carry out his divine will. The first being he made was "Michael," his "only begotten son." In the bible book of Proverbs, Chapter 8:22-30, we can read the account of the Son's creation in his own words. "*Yahweh* (The First Cause) himself produced me as the beginning of his way; The earliest of his achievements of long ago. From "time" indefinite I was installed, from the start, from times earlier than the earth. When there were no watery depths I was brought forth as with labor pains, when there were no springs heavily charged with water. Before the mountains themselves had been settled down, ahead of the hills I was brought forth as with labor pains: When as yet he had not made the earth and the open spaces and the first part of the dust masses of the productive land. When he was preparing the heavens I was there: When he decreed a circle upon the face of the watery deep: When he made firm the cloud masses above: When he caused the fountains of the watery deep to be strong: When he set for the sea his decree, that the waters themselves should not pass beyond his order: When he decreed the foundations of the earth: When I came to be beside him as a master worker: and I came to be the one he was specially proud of day by day, I being glad before him all the time."

In the Hebrew Bible Book of Genesis we can find a detailed record of the sons of God, the angels, having visited the earth many thousands of years ago, reading from Chapter

6: "Now it came about, that when men started to increase in numbers on the surface of the ground, and daughters were born to them, then the sons of the true God (angels) began to notice the daughters of men, that they were good-looking, and they (spirit beings) went taking wives for themselves, namely, all whom they chose." Here we find recorded proof that spirit beings not only came down to the earth, but were taking fleshly wives of men, even having offspring by them. This mixing of spirit and flesh was a detestable act in *Yahweh's* eyes. Therefore, the "First Cause" made a far-reaching pronouncement. He warned man that he would bring total destruction upon the earth after 120 years. He saw that the earth was full of wickedness, and he would bring all life to an end, sparing only eight souls (people). The sons of God he condemned to dense darkness in a place called Tartarus. These wicked angels who chose to forsake their spirit existence in the heavens were no longer privileged to remain in *Yahweh's* spirit realm.

Is it not clear that alien beings came to earth by their own choice to mingle with fleshly beings? If alien beings had the technological capability to transcend time and space

11

thousands of years ago, would it not be reasonable to assume that the favored angels of *Yahweh* would have an even greater technical capability in our own age? Anyone who would deny this undeniable reality has self-imposed blinders over his eyes of understanding.

Those spirit creatures condemned to Tartarus also identified as "demons" were, and are under the influence of "Lucifer," or Satan, who is the "resister" to *Yahweh* and his purposes. Satan and his followers, both demonic and human will do every thing within their "limited" power to disrupt the soon to come transformation of a corrupt, wicked, immoral, rebellious world system of religious and political domination. Lucifer will in fact, marshall all his angelic demonic forces against *Yahweh's* own discretionary declaration to establish his righteous kingdom and rulership over the heavens and the earth.

Clearly, man does not exist alone in this vast complex universe. Mankind is in fact; alone only in the sense that man has been isolated from other created beings because of his continued rebellious attitude. His isolation will continue until the First Cause cleans the earth, as in Noah's day, of all satanic evil. It is only after Satan and his world system of evil men have been destroyed that the earth will be readmitted back into the eternal kingdom of God.

You may wish to read this account from the book of Revelation, Chapter 14:9, where John said; (angel speaking); "If anyone worships (supports) the wild beast, and its image (United Nations) he will come under the wrath of *Yahweh*." Also we find recorded in Chapter seven this statement; "I saw four angels standing upon the earth holding tight the four winds of the earth, and another "angel" came from the east (sun rising) having a "seal" of God (the First Cause) saying, do not harm the earth or anything until "we" have sealed the slaves of "our" God in their foreheads". The number that was sealed (chosen for heavenly existence as immortal spirit beings) was one hundred and forty-four thousand. These chosen ones are the "elect", or "anointed" ones of *Yahweh*, and will reign with "Michael" the "King" for one thousand years while the dead in "Christ" are being brought back to life in the promised resurrection.

This fantastic transformation of the world of mankind will be carried out by thousands of angelic beings of *Yahweh*. The work will be directed by the king son, Christ Jesus, and his associated kings, the one hundred and forty-four thousand redeemed from the world of mankind.

Again, we find recorded verification of alien spirit creatures being sent to the planet earth to full fill *Yahweh's* purpose in the restoration of mankind. The "four winds" of the earth represents the natural elemental forces of the planet's ecosystem, storms, lightning, earthquakes and oceanic forces. All those forces will be brought to bear against mankind's evil system in the cleansing and restoration of planet earth to its original garden paradise condition. As shown, the five angels mentioned are commissioned to spearhead the transitional period. Note too, that the five angels were "sent" to the earth for a specific purpose at which time they are at this very period doing the marking work of *Yahweh's* slave-like ones.

In the fulfillment of his will and divine purposes, the grand creator has no concern or interest in man's viewpoint as to his belief that the angels are now accomplishing this work. He has said it, and it will be done, as *Yahweh* "cannot lie." His pronouncement made thousands of years ago is now being executed, and no power of combined force on the earth can foil or shunt that attainment.

CHAPTER 3

EARTH'S VISITORS

If we could regress two thousand years of man's existence on planet earth, we would still find recorded historical events involving alien beings having visited our earth on hundreds of occasions. Most people unknowingly pass these events off as myth or fable.

We have indisputable evidence that will verify beyond any possible doubt, that we have been under alien surveillance since man's beginning. Denial or subliminal fear will seem to threaten many by a new truth they cannot comprehend. They now realize they must make a decision. They must now accept this newly revealed truth, or they must reject this new truth and continue to hold on to an illusion realizing the immediate consequences of their choice of lifestyle. This reaction is a direct result of the implications associated with accepting this new difficult to perceive knowledge of the alien reality.

The powerful evidence of alien being visitation as revealed in the inspired words given to mankind by the First Cause is not to be viewed as myth or taken lightly. We can be fully confident that the information that "Elohim" provided in reference to his spirit creatures, the angels, he was giving us overwhelming proof of their undeniable existence. An example of this can be found in the very first book of his word at Genesis, Chapter 19, where he dispatched three of his angelic beings to destroy the two cities of Sodom and Gomorra. The appointed angels were under direct command to warn Lot (who found favor in *Yahweh's* eyes), and his family to flee the entire district, as they were soon to destroy the sinful cities, which they did. This recorded incident of destruction cannot be disputed, as the sites of those two cities are known even today. The incinerated remains have been located and partly excavated by archaeologists in recent times.

The method of destruction used by the angelic executors could have been a form of nuclear energy, as fused sand turned to glass has been found throughout the area of destruction. Fused glass is a common by-product of a nuclear detonation. This indicates temperatures as high as ten million degrees produced in the fiery destruction of the two cities. Is it not possible that *Yahweh* used the same energy we see in the nuclear fusion process in the sun and billions of enormous stars he created throughout the universe? How much more "proof" do we need?

Scientific skeptics of the angelic alien existence still clings to the unprovable misleading theory of blind evolution. The imposed teaching of this explanation of design, selection, and the bringing into existence of "all things" by pure happenstance is in fact, in the same category of deception as the lie Satan used to deceive Eve. This deceptive theory, being pushed upon the young minds today, is a leading cause of destroyed faith and belief in the creation reality for millions of "border-line" believers over the past few generations. This evolutionary supposition can never overcome the numerical odds of life and substance coming into existence through an impossible means of "random selection."

Such an unprovable, illogical supposition (or conjecture) would require considerably more "faith" to accept or believe as "fact" than that of accepting visible proof of the creation reality, as stated in the Biblical record, as provided by the "First Cause"; *Yahweh*.

As designer and maker of all things existing, both visible and invisible, he and his first created "son" Michael, knew long before mankind's rejection of the divine reality that man would rebel and deny that he, through his son, had formed all things.

This very attitude of man was made known to mankind, and has proven to be so as he recorded the following in the Bible Book of Ecclesiastes where he said, "And I got to see all my works that I had made, and how mankind (scientific establishment) is not able to find out the work that has been made under the sun. No matter how hard man seeks to find out, no, they do not find out. (Speaking of those who try to understand how things came into existence), and even if they (man) say they are wise enough to find out, (the works of the true God) and say they know, they are unable to find out."

Here we see that "The Wise One" knew that "His" wisdom, knowledge, and technical expertise was so far advanced over man's thinking capability that he could never understand how God's works were accomplished no matter how many ideas, studies, or theories he developed, he could never gain the "deep works" of *Yahweh*. Just one more example of how limited man's knowledge and understanding really is when compared to the wisdom and complexity involved in the creative process. His word tells us in great detail how he brought all things into existence, both the visible and the invisible, while man's feeble attempt to explain this enormously complicated "works", as having been caused by a chance evolutionary interaction between nucleic acids (DNA) and protein molecules (RNA). They never attempt to explain where the DNA, RNA, and thousands of the other "elements" of non-living matter came from. The process required to allow "life" to come into existence through an evolutionary "chance" concentration of a "swamp broth soup" of prebiotic content is absolutely moronic. *Yahweh's* creative accomplishments leave no room for ignorant speculation.

Another example, which proves to all that angelic extra-terrestrials have in the past, and are even now, visiting our earth is found in the book Ezekiel recorded for us. He relates his "first person" encounter with four alien creatures sent to him with an instructional message from *Yahweh*. Ezekiel gives us a fully detailed account of his visitors and the spaceships they had came down to earth in. His clear description of their spacecraft as "a wheel in a wheel with lights (eyes) all around" is an unmistakable distinct report of what is known today as an "unidentified flying object" (UFO). No doubt, Ezekiel certainly identified them.

Ezekiel knew they were spirit beings, angels being sent to him by the First Cause for a predetermined purpose. His detailed observation of the craft as having eyes all around was his way of describing "lights" around the perimeter of the saucer-shaped flying vehicle. There are many such records of alien being visitation to earth, both historical and modern sightings, if one wishes to research the subject. But remember, direct person to alien communication today is rare, and is only possible where such contact is necessary to fulfill the divine commission "they" have been sent to carry out, in most instances. Even then, the contact will be fully under "their" control. If you are directly contacted, it will be at a "place" and time they choose, and you will have no control over the proceedings during the contact. In ninety-six percent of the incidents I investigated where even near contact was involved, the person or persons implicated, said the encounter left them unsure of what they had just seen. Some said they even had difficulty recalling the exact details of the incident.

This full control situation is under the direction and influence of the First Cause, and cannot be altered in any way, unless it is within the scope of his will to do so. It is to be only after this planet has undergone the soon-to-come transformation, and reclaimed back into the universal brotherhood that mankind will be accredited the divine privilege of communicating with other created beings throughout the universe. *Yahweh*, the First Cause, told mankind that even the angelic beings in the heavens themselves will "rejoice" when the earth is "reclaimed", as at the "time of creation."

Although angels in their spirit form are invisible to human vision, as the wind is invisible, God, in the creation of his spirit beings, endowed them with exceptional abilities. One of which was their capability to materialize into human-like (visible) form when appearing to men. They were not in their natural state. The angelic beings are extremely active, have eternal life, are holy, and are able to demonstrate great force. Their super-human characteristics of angelic entities are not limited by the physical or celestial "laws" imposed on mankind, such as: gravity, time, distance, need for food and oxygen, etc. They have "total control" over all the physical and universal statutes, including energy in all its forms. In other words, they have attained complete technological perfection, to the degree that "nothing" is impossible for them, where it is within the "will and purpose" of *Yahweh*.

CHAPTER 4

ALIEN MISSION

Some very stimulating questions relating to alien earth visitation to all areas of the globe are: Why are they coming here? Who are they? How long have they been observing our planet? Why are they so evasive? Such mystical questions have baffled and confounded mankind for many, many years now. However, the answers to the questions are not all that difficult to find if we would but search in the source of divine truth. The confusion enters into the matter when we refuse to acknowledge and accept the obvious truth when we find it. For instance, our grand creator, *Yahweh*, told us to search for "accurate knowledge." So, where do we find "accurate knowledge"? The solid answers to the above questions can be found in only one place; His "word," found in the sixty-six books that make up the "Holy Bible."

One of the first recorded instances of alien visitation to earth is to be found in the first book of the Bible, at Genesis, where we read of man's first transgression against his creator *Yahweh*.

Here we are informed that the "first" woman, Eve, had been deceived by a serpent. Now, most people who read this would immediately conclude that a serpent couldn't even talk, so how could it have misled Eve? This is the very first mistake in our reasoning. *Yahweh* had the account recorded for us. And since he said' "I cannot lie," and, "all scripture is inspired by *Yahweh*, and is beneficial for teaching, reproving, and setting things straight." We must take his word (all of his word) as truth. If we doubt, question, or test his word, we are right then repeating the same mistake that Eve committed when she reasoned with the serpent and questioned God's instructions. Her actions led to death. The serpent which "spoke" to Eve was in fact a spirit being who had rebelled against *Yahweh*, using the serpent by transferring his spiritual powers to the "form" of a serpent. We should always be aware, that alien beings (angels) have the superhuman ability to transform themselves into anything they wish to become. This capability is extremely difficult for human beings to comprehend. This being so, most will choose the thing they can understand, which is usually a "half" truth, or a lie, as this is easier to believe.

The spirit being, Satan no doubt chose Eve to deceive because he knew that Eve was the weaker vessel, as she had been "taken" from man. A point to consider here is that a spirit being (an angel) was here on earth having direct communication and influence on the thinking and actions of a human creature, Eve. This influence over mankind has resulted in a woeful existence for mankind ever since.

After Adam and Eve had rejected the direct instructions given to them at the time he placed them in the garden paradise, God banned them to a life of pain, labor, and finally death. After which he posted his spirit angels, called "cherubs," at the east of the garden to prevent their reentering. His written word tells us, "And so, he drove man out of the garden of Eden, and posted at the east of the garden entrance to keep them from reentering, a flaming blade of a sword, that was turning itself continually to guard the way to the tree of life." Note that all this happened right here on earth. The angelic beings had been sent to earth for a specific divine purpose. Their assigned mission was to guard the "tree of life."

In every instance where *Yahweh* had dispatched his spirit beings to planet earth, he had given them clear assignments he wished to be accomplished. Their assigned mission never failed. His angelic warriors could not fail because they came with *Yahweh's* backing, authority, and unlimited power.

By way of a reminder, listed are "just some" of the paranormal powers and abilities the spirit sons of God possess:

1.) They are given full freedom to travel to any part of the universe, earth included.

2.) They are not limited, controlled, or under the influence of time.

3.) They see the future course in time.

4.) They have an eternal existence.

5.) They are sustained by energy infusion. (Everything in the universe is a form of energy.)

6.) They are able to exist, and transform, or materialize, into any form or shape, physical, material, ethereal or energy force form.

7.) Ability to transmigrate to any point in the universe instantly.

8.) They possess unlimited power and can "bend" universal "laws."

9.) They possess unlimited love.

10.) They have been appointed "guardians of universal creations," man included.

All the above listed paranormal abilities emanate from the first cause, *Yahweh*, the only true God. This is but a small listing of the super-natural capabilities of spirit aliens. *Yahweh's* angels are not restrained, but are "all powerful."

You may recall that Jacob, the son of Isaac, Being one of *Yahweh's* faithful servants on the earth had a face-to-face encounter with one of Gods angelic heavenly beings. The angels' visit with Jacob is found recorded in the thirty-second Chapter of the Book of Genesis, where we read, "As Jacob was on his way, angels of *Yahweh* encountered him, and when Jacob saw them he said an encampment of *Yahweh* it is. And he named the place, "Place of two encampments." This close Encounter of real alien beings was not fiction or myth. The ancient writer of the book of Genesis goes on to relate an amazing detailed description of Jacob's visit with the divine beings, "And Jacob was left alone; and there with him a "man" wrestled with him until daybreak. And seeing that he could not prevail over Jacob, he struck against the hollow of Jacob's thigh, and Jacob's thigh was dislocated as he wrestled with him. And the angel said, "Let go, for it is daybreak." But Jacob said, "I will not let you leave unless you bless me." And the angel asked him, "What is your name?" And he said, "Jacob." And the angel said to him, "Your name shall no longer be called Jacob, but rather "Israel," for you have striven with both *Yahweh* and men, and prevailed." And Jacob asked him, saying, "Do tell me your name!" And the angel said, "Why do you ask my name?" And he blessed him there."

This fantastic incident with Jacob clearly shows that men of old who gained *Yahweh's* favor could have an extra-ordinary relationship with divine alien angelic beings.

Abraham had similar first person encounters with *Yahweh's* spirit creatures. You may remember that an angel of the true God appeared before Sarah, Abraham's wife, and told her that even in her advanced years she would bear a son, who was to be named Isaac.

Later, Abraham himself was visited by three angels. The three angelic emissaries were "sent" to planet earth to make an inspection of the cities of Sodom and Gomorrah, the purpose of which was to verify the extent of sinfulness before bringing the cities and inhabitance to total destruction. Having confirmed *Yahweh's* judgment as justified, they did indeed fulfill their mission, and destroyed the cities. This fiery destruction was accomplished by two of the angels while the third angel remained with Abraham.

Would it not be reasonable to suppose that if alien beings visited our planet earth on a regular basis over hundreds of years in the past, that this alien visitation is being carried on even now? Yes, it is not only reasonable, it is observed fact. The only difference today is now we call the visitors UFOs, or unidentified flying objects, when in fact they are not unidentified, but unadmitted flying objects.

The thousands of unusual sightings reported every year cannot be explained away as misidentification by the observer or propagandizing the general public into accepting the idea that an observer of this phenomenon is some psychological misfit within our society. To be sure, the experience can, and does have a psychic impact on an observer. However, this normal expectation in responding to an unusual event does not justify classifying an observer as unbalanced mentally or afflicted by hallucinations. The systematic deliberate promotion of this viewpoint among society has resulted in thousands of UFO (IFO, identified) observers deciding to keep silent about their sighting experience. Out of more than a score of the people who reported to me as having personally seen these strange crafts, less than four percent told me they had reported the event to local law enforcement, or otherwise mentioned their sighting to anyone outside the family, or close friend circle. If this is a predominate reaction among society in general, this would indicate possible hundreds of thousands of visiting alien craft observations going unreported and therefore being "unknown" to all-except the observer and his immediate circle of confidants. One reason for the reluctance "to tell", could very well emanate from the spirit endowed angelic occupant in the craft, as such beings are known to possess super-natural psychic influence over the human mind, as in the days of the prophets.

Some may speculate that the UFO occupants may be "demonic spirits." This assumption does not agree with recorded physical reality. At the time of the Noachian flood, the rebellious demonic angels were no longer allowed by *Yahweh* to materialize into the physical state. They were limited to the spiritual realm (which is invisible). The spacecraft (UFOs) are "visible" to the human eye. This is proof that the vehicles are operated by spirit angelic creatures remaining in *Yahweh's* favor, not "disfavored demonic spirit beings". The "Holy" beings of the spirit realm still retain the ability to materialize into the physical state (visible to man).

Clearly, the space chariots (UFOs) and the occupants are in the visible state (physical) as they are being seen and reported all around the globe. This being true then, it would verify that the crafts being seen are without question angelic creatures under direct authority of *Yahweh*, the "first cause," and not demonic spirits.

Reading from the Bible Book of Revelation, Chapter Seven, we will see that the four angelic beings sent to the earth by *Yahweh*, were given the same mission that the three angels sent to Sodom and Gomorrah were given. *Yahweh* speaks, "Do not harm the earth or the sea or the trees, until we have sealed my slaves in their foreheads." The harming of the earth (the entire earth) by the appointed angels parallels the like destruction visited upon Sodom and Gomorrah by two angels sent. However, they had instructions from *Yahweh* not to harm the cities until his "righteous ones" had fled the district. Soon after Lot and his family had reached safety in the mountains, the cities were made desolate. The four angel warriors identified in Revelation are now awaiting the "sealing" of his slaves before the

destruction of the modern day Sodom and
Gomorrah, which is the current "wicked
system of things". This destruction will involve
the armies of the earth, battling with "Michael"
and his army's war chariots.

CHAPTER 5

NO FAILURES HERE

The year 1973 was known as "the year of the aliens." That year recorded more unusual spacecraft (IFOs) than any year prior to, or since. Having no means of communicating with the occupants of these super-natural alien visitors, one can only speculate as to "why" 1973 was a banner year for earth visitation by angelic beings. But wait, is it really speculation? No, those who would speculate this issue are merely engaging in a risky mental venture to fill in the blanks where true knowledge and reality is denied, or lacking.

You can be absolutely sure of one truth. They are not coming to this earth on a sightseeing tour. The grand creator of the celestial bodies is personally interested in the earth. He will very soon now, demonstrate his unchallengeable sovereignty over this unique planet. *Yahweh* has warned mankind for thousands of years now, that He and his highest of all the celestial beings, "Michael," will restore complete righteousness over the

35

world of mankind. That restoration is undergoing preliminary preparation in this, the "final days," of this worldly system of things.

The expression, "world system of things," is to be found more than thirty times in the Greek scriptures. The term relates to the current condition of world existence, and the totality of world affairs, which manifest itself in the course of time, or period of domination by man. We may speak of the world system of things as an "age," "era," or "epoch." It is also expressed as "this system of things," "this wicked system of things," and "this present system."

At Galatians 1:4, the apostle Paul wrote, "He (Jesus) gave himself for "our sins" that he might deliver us from the "present wicked system of things," according to the will of our father, *Yahweh*." Paul goes on to warn us not to be fashioned after this present wicked system, which is controlled by demons (spirits) and evil men, all of which will, soon be destroyed. *Yahweh* identified the "GOD" of this wicked worldly system of things as none other than "Satan," the rebellious spirit angel.

In this, your own generation, you are an eye witness to the pre-work *Yahweh's* angelic beings are doing all around the earth to bring the world of mankind into direct notification of the coming destruction. An angel (Rev. 14:6-7) speaking of our time, "the last days" said, "Fear *Yahweh*, because the time of (for) judgment has arrived." Note that this warning was delivered to "the earth."

Human governments or political leaders cannot solve the cascading problems confronting the world of mankind today. "Critical times" are impacting every living being on this planet. This condition was seen long ago by the apostle Paul. Where he wrote in his letter to Timothy at, Second Timothy, Chap. 3:1-5. He said, "In the last days (of the present world system) critical times, hard to deal with will be here." He tells us why. He reveals, "Men will be lovers of themselves, lovers of money, self-assuming, haughty, blasphemers, disobedient to parents, unthankful, disloyal, having no natural affection, not open to agreement, slanderers, without self-control, no love for goodness, betrayers, headstrong, full of pride, lovers of pleasures rather than lovers of God, having a form of Godly worship, but denying the power of God. From such ones we should flee." Is it not clear to all, that all evil and wickedness in the earth today spring from the things Paul listed here? The list Paul gives us here is clearly the "traps" Satan has employed for thousands of years to ensnare mankind. All those who have been "caught" in Satan's traps are to be counted among the "many" that will experience *Yahweh's* retribution. His anger will be poured out upon all those who have "Satan's mark", at the final battle to be fought between the "powers" of the earth, and the angelic warriors of *Yahweh* at "Armageddon", a war of which the victorious has already been pre-determined. (Read Daniel, Chap. 2:44). Here the victor was revealed to Daniel by an angelic messenger over two thousand years ago. The visiting angel said to Daniel, "Now in the days of those kings, the God of heaven will set up a kingdom that will never be brought to an end. And this kingdom will not be passed on to any other people. It will crush, and put an end to all earthly governments and kingdoms, and it will stand to times indefinite, even forever."

This everlasting kingdom transition is to occur in "the time of the end," or last days of this world's system of things, the period of time in which we are now living. This very fact explains why the celestial angelic warriors are now coming down to the earth in ever increasing numbers. As the "last days" progress nearer to their end, *Yahweh's* angels will increasingly become more assertive in making themselves known particularly among the ruling element of the earth.

Occasionally you will hear a spiritually blind world leader make reference to the coming "new world order." This grand utopian world system can never be brought about by man, and will therefore never be realized. Why? Because man's idea and hope for a utopian world society is to be based on a man-made foundation of greed, and man's desire for total control over the riches of the world. This includes labor, resources, finance, and rule.

To be sure, the only true utopian society ever to be established here on this earth will be by and through the heavenly armies of *Yahweh*, the God of creation. He makes this plain to us in the last book he had recorded by his servant John. In Revelations, Chapter 21 he wrote: "And I, John, saw a new heaven and a new earth, for the former heaven and the former earth had passed away, and the sea is no more. I saw also, the Holy City, New Jerusalem, coming down from God, prepared as a bride adorned for her husband. With that I heard a loud voice from heaven say, "Look, the tent of God is with mankind, and he (*Yahweh*) will reside with them." (Here on the earth).

This heavenly kingdom over the earth represents the true utopia. This divinely led government is not founded on greed, or animalistic evil scheme, but will be based upon love, joy, peace, long-suffering, kindness, goodness, faith, mildness, and self-control. Against these things no law is needed.

The disciple Matthew warned us that in this time of the end, world conditions would progress in wickedness to a point where mankind would literally destroy all "flesh" on the planet earth. But *Yahweh* said he would intervene in man's affairs to prevent this catastrophic suicide of all "flesh." He does not intervene to save mankind as a whole, but does so to "protect his chosen ones". Again, this intercession will be forcefully imposed upon mankind by God's war chariots, which are manned by angelic warriors. Is this to be a difficult mission for his angelic warriors? No, precedence was set in the days of King Hezekiah's war with King Sennacherib. King Sennacherib's army (Assyria) had completely surrounded the city of Jerusalem, when *Yahweh* sent one angelic warrior to King Hezekiah's aid. The angel's power was well demonstrated in one night when he (the angel) killed one hundred and eighty five thousand men. Does this example of the power displayed by one angel in one night leave any doubt as to the battle of Armageddon? Should we question the ability of thousands of angels to overcome the combined armies of the nations soon to be confronted on earth?

CHAPTER6

MAN'S RESPONSE

Ever since the creation of mankind in the year 4027 BCE, man has remained the highest order of intelligence among the living creatures. Even so, man's intellectual capacity in understanding the massive "creative works" of *Yahweh's* divine activities is narrowly limited. Although he, in his never-ending fire of curiosity, eagerly searches for wisdom and knowledge seen in the design and fashioning of the awe-inspiring universe.

The scientific community will not submit to the realities seen in the creation. Instead, science seeks to "explain" how the creative works came into existence. They make every attempt to belittle or disprove the creation reality through unprovable, ambiguous theories. Not one "scientific" theory in attempting to explain the "creation," can override the simple Biblical statement, "In the beginning, God created the heavens and the earth." *Yahweh* did not intend to reveal to man how he formed the universe. He knew at the beginning that man did not possess the mental faculties to perceive such advanced wisdom. Therefore, man, in his most advanced stage of ignorance; will continue to develop his unprovable suppositions.

Where mankind finds himself lacking in true wisdom and understanding, he will attempt to mask this mental limitation by formulating theories of denial. His theories and suppositions may appear plausible in his own understanding, but is in fact, an attack on *Yahweh's* word and sovereignty. In most situations where man is incapable of gaining true insight as to reality, he will invent his own reality to conceal his ignorance.

The UFO reality is an example of just how far the scientific establishment is willing to go in its effort to conceal the truth. Every responsible level of social influence has been strangely silent in explaining the UFO phenomenon. Why? Over the years, in my search for "real answers" to this unresolved enigma, I have interviewed more than a score of honest, down to earth, trustworthy people, who have been surprised observers of visiting aliens of celestial origin. In every visual encounter, the observers were absolutely sure that what they had seen was something they had never witnessed before. Eye witness attestation came from people of every level of society, including, school teacher, farmer, professional, aircraft pilots, Air Force aircrew members, and university students. These people not only seen the never-before-seen spacecraft, but were able to recall specific details of the event. They were not lying about their experience.

A large painting (24" x 18") of an extra-terrestrial, triangular shaped spacecraft is on display at the science research museum at Roswell, New Mexico, of one of the craft seen near Statesville, North Carolina. The craft had been observed in the same area around Lake Norman, N.C., by six different known observers. All were found to be sincere, creditable eyewitnesses.

After having submitted UFO sighting reports to the local news outlets, not one report has been aired or publicized by any media source. Why? Could it be that some, in the higher order of society, have seen reason to silence all major news outlets regarding UFO activity? Or is the political world encouraging silence to postpone an upheaval among the general population? You can be sure the "silence" blackout is a worldwide conspiracy with the scientific community leading the pack. The cover-up is fundamentally of interest to the scientific field because they refuse to admit they exist because science has no answer as to what they are, or where they come from. That is why the scientific establishment will go to almost any extreme to disclaim any report that does slip through the news media screening network.

An interesting oddity associated with almost every UFO sighting I investigated was the observer reported an "unusual" spiritually psychic experience, which lingered for some time after the encounter. This extraordinary consequence has been a commonly reported effect by thousands of UFO observers from all over the earth. No unexpected surprise, in the way of a reminder, the UFOs (craft) are under the control of spirit beings (angels), which could (does) radiate "spiritual energy" to the subconscious, which would have a spiritual influence on "man's psyche". Their powerful spiritual energy has an immediate impression on the mind and senses. No two-way communication is necessary to instill this effect, yet the effect may persist for years. In some incidents, this "psychic shock" has even changed the human personality. Could this awakening of spiritual awareness suggest religious connotation? Yes, *Yahweh*, in his flawless wisdom, will not permit mankind to continue for long on his present course. He informed mankind long ago in his written word, that the "spirit searches" all things. Man will shudder in fear when he sends his angelic warriors to make their final examination of the planet earth in surprising numbers to intercede in world affairs. World leaders will see this as an alien intrusion, and will unite all their military forces into a "one world army." This combined military assembly will unite to oppose this "alien" (angelic) threat. This worldwide course of resistance will set the stage for the last war ever to be fought on planet earth.

This war of "God the almighty" was foreordained by *Yahweh* in the book of Revelation. All those who (on earth) support the opposition forces against *Yahweh's* heavenly armies will be annihilated at Armageddon.

The conclusion of this war of execution will usher in a time of peace and the beginning of the "new kingdom" over the refined "new earth", as shown in the book of Daniel, Chap. 2:44. It is at this time that "little baby Jesus" (heavenly name, Michael), becomes "commander in chief" of the heavenly war chariots, which are now victorious over all opposition in the heavens and the earth, in the most superlative degree. Right up to the beginning of the battle of Armageddon, world political leaders were looking forward to a time of scientific progress and technological wonder. Their vision of a one-world government would result in a utopian society under a "new world order." However, just the opposite was true. Our present generation has seen the fulfillment of the signs of the "end time" given by Jesus in the twenty-fourth Chapter, from the Book of Matthew. With the fulfillment of these "signs," the next earth-shaking event will be the conclusion of "this system of things." From the human standpoint, the turning point of world history will be the total destruction of worldwide false religion.

No, the future of mankind will not be determined by his own making through science breakthroughs, or some abstract evolutionary process initiated by happenstance, no mankind's future lies solely in the hands of *Yahweh's* super beings, the holy angelic host. If we, as created free moral agents, want to live forever under perfect world peace and order, we, as free agents, have been given the freedom to make that choice. True, there are many that will refuse to accept and conform to *Yahweh's* clear standards. For those, the resulting consequence is a result of their own failure, and so, is it not wise to follow *Yahweh'*s counsel when he urges us to. "Take warning before there comes upon you the day of *Yahweh's* anger, seek him all you meek ones of the earth. Seek his righteousness, perhaps you may be concealed in the day of his anger." Are we proving ourselves to be meek of heart? The spirit of *Yahweh* searches "all things."

CHAPTER 7

A REASON FOR HOPE

Under man's present system of rulership, the world of mankind has little hope of true happiness. The greater number of people on earth has labored through their entire short life searching for something better. Even so, few have ever attained true love, joy, peace, goodness, mildness, or self-control. The inability of mankind to gain these highly desirable fruitages can be attributed to mankind's persistent rejection of true knowledge and wisdom. Michael, the only begotten son of *Yahweh*, said, "This means everlasting life, the taking in "accurate knowledge" of *Yahweh*, the father, and the one sent forth." (Jesus Christ).

The real key to finding true happiness and life eternal is based upon our desire to really search for the wisdom of God, which desire comes from his accurate knowledge of the truth. Where do we find this unequaled knowledge? In this modern technological revolution in the scientific world, it is really

49

difficult, but is it wise to put our faith and trust in man or his system? No, the historical record shows that ever since *Yahweh* allowed man to govern himself, mankind has sunk deeper and deeper into the abyss of wickedness and spiritual darkness. When we look at conditions in the earth today, is it not clear to all, that man, in his own way, does not possess the divine wisdom, knowledge, and understanding to even direct his own footsteps? *Yahweh* warned us long ago. And he said, "There is a way that "seems right" to man, but the end thereof is death." Surely we would never put our faith or trust in one who is leading us to death.

This being true then, where can we turn to find the "direction" needed to lead us to life instead of death? There is only one source of life saving wisdom and knowledge available to mankind. That source is freely given by our creator, *Yahweh*, who represents "all" wisdom, knowledge, and understanding.

How would you feel if you were an artist and had just finished a masterpiece you had been working on for years, and you gave it to the local art museum as a loan, to care for, and when you came by to check on your creation, you found that the curator had ruined it because he no longer trusted you? How would you respond to this lack of appreciation? No doubt you would be hurt, would you not?

Well *Yahweh*, the master creative artist, designed, and with great wisdom, fashioned an unmatchable masterpiece when he brought the universe into existence. He gave a part of his creation, the earth, to the newly formed family of mankind. He asked them to care for their new paradise home. He entrusted them to take care of all he had made on the earth. He had envisioned a happy, glorious, everlasting future for mankind. All too soon, man lost his trust in *Yahweh*, their creator, and no longer believed what God had told them. Why? Because one of the foremost angelic beings came to the earth, and in the form of a serpent, persuaded Eve (the first female human), to rebel against *Yahweh*. This (by choice) rebellion grieved *Yahweh* very deeply. It was at this very moment in man's existence, that he lost God's favor and wisdom, and has never regained that wisdom of God, even though *Yahweh* has given man over six thousand years to do so. Instead, mankind has chosen to go his own way in rebellion and denial. His wisdom is still available to all that sincerely desire to come to know the true God. His wisdom can be found in his word, the Holy Bible.

In the soon-to-come transformation of planet earth, the "real life," everlasting life, will be given to all those who possess "the wisdom from above." As revealed earlier, *Yahweh* will destroy all those who share in the current "wicked system of things", as the world of mankind is rapidly nearing the end of "the last days" which began around 1914, at the end of the "gentile times." You may remember that the "last days" before the Noachian flood as shown in Genesis, lasted 120 years before the destruction of all living things, (except those spared). If the "last days" of our own age were in the same range of time, we would have but a few short years remaining before our own period of transformation comes. This can be verified by Biblical prophecy.

The transformation of planet earth will bear upon all mankind, giving "life," or "death." *Yahweh*, being the merciful God that he is, has reminded us through his angelic spirit beings, that; "He is patient because it is not his desire that any be destroyed, but desires all to come to an accurate knowledge of the truth and remain alive." Herein lays our only hope for a future life without end. We cannot buy, bribe, or lie our way into life eternal. It is a gift of *Yahweh* for demonstrating our faith, trust, and love for him and his mighty son, Christ Jesus. The only key to finding eternal life is found in his word, the Bible. To deny this truth is to condemn ourselves as well as others to die. (Read James, 5:19).

Most people have very little understanding of what death really is. Many have been erroneously led to believe that when we die, we are not totally dead, but are immediately judged, and are then sent to either "heaven or hell". Nothing can be more deceptive. This concept does not agree with *Yahweh's* own word. Most people alive on earth today are not aware that all living human creations are subject to "two deaths." How is this so? The word of God shows that, the first death we suffer is a result of our sinful state. We inherited our first death from our forefather, Adam, the first fleshly being. *Yahweh* warned the first family; "The wages sin pays is death."

This first death, all men will suffer, because, "All men have sinned, and fall short of God's glory." All who are in the state of the "first' death are neither in heavenly bliss, nor are we thrown into a tormenting "hell", to burn forever, as many have come to believe. This is because of false doctrinal teachings by apostate teachers. No, in the condition of the "first" death, we are placed in a state of prolonged rest (sleep), to wait the day we are to be resurrected, as Jesus himself was resurrected. From the Bible Book of Revelation, Chapter 20, you will read of the first and second resurrection period. You will find that those having part in the "first" resurrection will become kings and priests of God, and will not be subject to the "second" death.

The second (general) resurrection will take place for the "rest" of the dead during the 1000-year rule of Christ, and his associate kings. It is after this 1000-year reign that those who were in the second resurrection will, at the end of the one thousand-year rule, be subject to the "second death." (See Rev. 21:8).

Since the resurrection promised by *Yahweh* will include both the righteous and the unrighteous, does it seem reasonable to "resurrect" the dead when, (according to apostate teachings) after they had already been sent to heaven or hell? Why would we be resurrected from heaven? No, those who die the first death are waiting to be resurrected from the grave. (Hell, Hades, Sheol; means mankind's common grave, not eternal torment.)

Such a teaching or belief would make a liar out of *Yahweh*, and his son Michael. Why? Because this is a doctrinal teaching of man, not that of our divine creator. (Read Acts, 24:15 in your own Bible.) No, this first death is in reality a condition in which God places us for an indefinite period awaiting the resurrection transition back from death to life. The 1000-year rule of spirit beings will mark the final test of mankind's loyalty to *Yahweh*. If we fail this final test, by Satan himself, it is then and only then, that we face the "second death." (Read Rev. 20:5, 7-9.) The most meaningful time of our life on earth is to come after the coming resurrection. This will be the period that determines our real hope for life eternal, or face the "second death," from which we will never return.

CHAPTER 8

HOW TO SURVIVE

The new earth kingdom paradise, to be installed in the very near future by *Yahweh*, will be under the loving rulership of the archangel Michael, and his 144,000 associate kings brought from the earth after being marked (selected) by *Yahweh*.

This kingdom will replace all world political systems, and no world government will ever rise to power to be ruled over by man. Bible prophecy warns all the inhabitants of earth that an unprecedented world change will soon transpire. This foretold event was given in detail by *Yahweh* in the book of Matthew, Chapter 24. Jesus, God's son, gave his disciples clear evidence of coming signs which would indicate the end of this present worldly system as we know it. Some of the major events he said would occur "in our generation" included:

1.) "Nation will rise against nation, and kingdom against kingdom." From the mid-1900's, this fulfillment is indisputable. There have been hundreds of wars among the nations over the past fifty years, and continues even today.

2.) "There will be food shortages." We are now seeing worldwide famine. As you read this, more than 400 million people worldwide are near starvation. More than 12 million children die of malnutrition annually.

3.) "*Earth*quakes in many places." The number of earthquakes being recorded all around the earth in our generation far exceeds anytime in record history. Over the last eighty years, hundreds of thousands of lives have been lost.

4.) "In one place after another pestilence." In 1918, more than 21 million people were killed by Spanish influenza. It was likely the greatest killer known to mankind. Other "new" and even unknown diseases have appeared all over the earth in the past few years, such as AIDS, Ebola, and many others, some are even immune to antibiotics, and are "quick kill" in effect.

5.) "There will be fearful sights." This has been called the century of fear. New weapons of warfare strikes terror in the minds and hearts of billions of people on earth. With the development of nuclear bombs, high fatality chemical, and biological agents, the total destruction of all life on earth is now in man's hand. This is an extremely fearful expectation. Increasing hostility among the nations is an ever-escalating cause for fear among the people.

Many agnostics would argue that these conditions have always plagued mankind from long ago, and are nothing new to man! They are quick to point out that these things have been happening for hundreds of years now. If this were nothing new, why would this be any different? The profound difference is, all the signs leading up to the end of this world's systems, are all being seen and felt in one generation: Our generation. Jesus, the Son of God, said, "When you see "all" these things occur in one generation, know that the "end" is near, even at the door." He continued saying, "I tell you, the generation seeing all these things will not pass away (die) until the end comes." You are living in that generation the "Angel Michael" was referring to here. We are seeing all these signs happening all around the world as they progressively get more intense, as the end draws near.

The neighborly love once shown by most people years ago, is a thing of the past. That love has been replaced by suspicion and distrust. One of the most defined "signs" that we are indeed living in the "end time," is the complete moral decay worldwide. Security and peace is no longer found in one's own home. Even so, *Yahweh* warned us that this very generation would see the birth of his heavenly kingdom to be ruled over by his spirit angelic son, Christ Jesus.

This kingdom was born in spite of unequaled world turmoil and distress. Since the year 1914, many historians refer to that year, as the year the world went crazy, and stands a turning point in the history of man. It was at that time that civilized man began a period of sick, terminal destruction of the human social structure.

Where do you stand amid this unrestrained, immoral society? We should do everything within our power to separate our family and ourselves from a dying world, being driven and controlled by Satan and his willing agents on earth. Jesus Christ, our teacher and leader, instructed us, "Do not love the world." If anyone loves the world, the love of *Yahweh* (God) is not in him, because everything in the world, the desire of the eyes (flesh), and the showy display of one's means of life (their riches and material possessions) does not originate with the father, but originates with the world. Furthermore, the world (evil mankind) is passing away, (to be destroyed), but the will of God remains forever. (The righteous ones left on earth).

Billions of people will reject these words of truth because of their desire to get all they can in their short life span. This very mental attitude will serve as a snare for them from which they are never able to escape. For such ones, little hope remains. No, many are not willing to sacrifice the "temporary pleasures" of this first life, for the divine pleasures of eternal in the coming new paradise earth. There will be millions more who will, maybe even agree that this is perhaps the truth, but soon return to their worldly routine, they too fail to act on the warning, and will suffer because of their greed. Then there are those God fearing "few" who will accept God's word as the truth, and will make an honest effort to please *Yahweh* by putting their full faith in his word. Their reward is everlasting life on a new earth. They will be protected through the horrors of the coming "great tribulation" soon to involve the entire inhabited earth, a tribulation that has already begun.

The "last" thing to occur before the end of the age, Jesus said, will be; "This good news of the (coming) kingdom will be preached in all the inhabited earth, for a witness to all the nations; and then, the end will come." This preaching work is now being carried out in more than 240 nations of the earth as you read this. This preaching work is being directed by angelic beings, and is in fact the separation work of the sheep-like people from the goat-like people, which Jesus spoke about in the book of Matthew, Chapter 25:31-35. He said, "When the son of man (Jesus) arrives in his glory (which has already happened), and all the angels with him, then, he will sit down on his glorious throne (in the heavens) and all the nations will be gathered before him (here on earth, not in heaven), and he will separate people one from another by bringing the good news to all, just as a shepherd separates the sheep from the goats." Jesus does not do this physically by pointing right or left, but by the word of the "good news." And he will put the sheep on his right (gives them his favor), and the goats on his left. These are rejected because the goat-like ones have refused to accept this "good news," and are therefore seen as goat-like ones. The "good news" of God's word is what does this separating of the people, and remember, the angels, (Jesus' helpers with his associates he "marked" on earth) are the means he uses to accomplish this work being done all over the globe. They are (the angelic beings) directing and aiding those on the earth by bearing the "good news" message.

The only hope of the true Christian, and indeed of mankind, lies in Jesus Christ. The granting of eternal life in heaven or on the earth was not possible for mankind until "Michael" was sent to the earth, to open the way of life. This is what our hope for eternal life is solidly based upon. It is supported by two facts. First, it is impossible for God to lie, namely, about his promises and his oath. And secondly, our hope that resides in Christ, who now has immortality in the spirit realm, the heavens.

With so many fulfilled prophecies in evidence today, the bible has without question established its truthfulness as the inspired word of *Yahweh*, the God who has shown mankind the end from the very beginning. He has shown by the infallibility of his word, that his word is not of men, but just as it truthfully is, the word of God. As such, you can put absolute confidence in the things he revealed to us. He said, "So my word that goes forth from my mouth will prove to be. It will not return to me without having accomplished my purpose. It will (His word) certainly do that in which I have delighted, and it will have certain success in that for which I have sent."

In our own generation, it is very enlightening to live in a time of man's history, when we are eyewitness of world events fulfilling God's prophecies as they happen in "the last days." This is proof that the time has come for his word to have certain success.

Thus, those who choose to be independent of the creator's purpose will be "cut off." While all those who are "hoping" in *Yahweh*, will live in the "new earth kingdom utopian paradise."

How many times have you heard the questions, "Why does man have to die?" and, "Can man die more than once?" The truth as shown in God's word tells us we may be subject to die two deaths. It is clear to all that we have been made subject to the first death through inherited sin from our forefathers. No one is exempt from this "first" death, as the wages of sin is death. Even though man was placed on the earth to live forever, his life has been placed in jeopardy by Satan's lie.

The penalty of the first death was all-inclusive in its effect until *Yahweh*, in all his wisdom, provided a way for man to regain life, even after death. At Psalm 68:20, we read, "To *Yahweh,* the sovereign Lord, belongs the ways out from death." As in case of "God's Holy Spirit," most people do not fully comprehend the condition of the dead. When one "dies", he is not in a state of immortality existing as a departed soul continuing life in an ethereal soul form. The word of God gives us a clear understanding of the state of death. King Solomon recorded at Ecclesiastes 9:10, "All that your hand finds to do, do it with your very power, for there is no work, nor devising, nor knowledge, nor wisdom in the grave (sheol), the place to which you are going." Does this suggest we have an immortal "soul" that continues life at our death? There is no such implication shown in God's word. The reality of death is; "We no longer exist." We return to the dust of the earth from which we were made.

Some are confused as to just what the human "soul" really is. The "soul" of man is the man himself. Man can exist in two forms as "a soul." He can exist as a "dead" soul, or a "live" soul, but never as an immortal soul. That is why Ezekiel said at 18:4-20, "The soul that is sinning, it itself will die." Clearly the soul is the "living person" before death, and the dead person after death. Nowhere does God's word say that man is an immortal living being (a soul).

So, death can be the final state of man. However, our grand God gave mankind a second chance for life in and through the shed blood of his only begotten son, Christ Jesus. This hope is true, and comes through a resurrection of the dead souls from the Adamic death. By suffering death, Jesus "tasted death for every man", and provided a ransom for all.

Those who want to be cleansed from sin's condemnation and released from the law of sin and death must allow themselves to be guided by God's spirit and deny the sinful inclinations of the flesh. This is much more difficult to do than it may sound.

Even though we may see the "first death," we have *Yahweh's* assured promise of resurrection at his appointed time. The coming resurrection will not only demonstrate *Yahweh's* unlimited power and wisdom, but also his unmatched love and mercy which will vindicate his universal sovereignty over Satan and his death dealing works. It is at that time death is swallowed up. He at the time also answers Satan's accusations that asserted that man would give everything he has in behalf of his soul (Life). Satan has been allowed (up until this time) to go the full limit in his deception of mankind. His false accusations have now been exposed as the "big lie." The suffering of man, his sickness, inhumanities, and all his evil works over the past 6000 years resulted from the overt act of turning Adam and Eve away from *Yahweh*, their God.

CHAPTER 9

MAKING THE WISE CHOICE

Most people seldomly think about living forever. Many feel that living forever is an impossibility, and still others never expect to do so. But would one be asking too much to want to live forever without dying? Looking at the past human condition, we would quickly realize that eternal life is well beyond man's own ability to attain. This being so, how can we reach a condition promising such life?

The first, and most basic step in laying a proper foundation for such a possibility of course is to accept the reality of Bible truth, while the second step is to bring our lives into total harmony with that truth. We can never come to know the "real life" available to us, by hanging on to this worlds beliefs, desires, or teachings. To do so only offers death as it's reward.

This difficult personal choice originates from the fact, that in our creator's wisdom, we were made to be "free, fleshly agents." We were given the freedom to choose good or evil, and where we wish to place our loyalty. As free agents, our individual choice determines our final fate: Life or death. Both will be ultimately eternal. Our creator, *Yahweh*, being all wise and all knowing, made some fantastic provisions to aid us in making a life giving choice.

In the Bible Book of John, we find that Jesus Christ, the one *Yahweh* sent from the spirit realm of the heavens, was called "His word." John revealed, "In the beginning, "the word" was, and the word was with God (*Yahweh*), and the word (Jesus), was made a God (the archangel). This one was in the beginning with God. (The first of spirit creations). By means of him was life, and the "life" was the light of the world. Here we are informed by John that Jesus was the life and the "light," or understanding of *Yahweh's* purpose would come to men through him, as he is God's spokesman. This "word" and "life" was given to man in the "flesh," which is now available to us in writing, in the inspired Holy Bible. It is only through a sincere diligent search of this written "word" that we come to gain the hope of eternal life. Because his word is eternal, any "life" gained through the "word" will also be eternal.

Jesus Christ, while in the flesh on earth said, "This means everlasting life, the taking in accurate knowledge." Getting the accurate knowledge leads to the wisdom of God, which in turn, leads to the right choice, which then leads to eternal life.

Why then did *Yahweh* make it possible for us to make the "wrong" choice? Well, our grand creator, from mans beginning, never intended that we make a wrong choice. This possibility came into being through our own decision as free agents. When the spirit creatures were created in the heavens, they too were made free agents. The scriptures indicate that the spirit creature known as "Beelzebub," (Satan) decided to exercise his free agency status and take a course of opposition and rebel against *Yahweh*. Jesus, another spirit son of God said of Satan, "That one was a man-slayer when he began, and he did not stand fast in the truth, because truth is not in him." His first overt act was in turning Adam and Eve against God by lying to Eve. This very act has brought the "first death" to all men. It is clear throughout the Bible record the qualities and influence he has gained over mankind can only be attributed to a powerful being, not an abstract principle of evil, or mythological apparition.

So, from a righteous, perfect beginning, this spirit being, as did man, chose rebellion, sin, and degradation. You may recall that Adam and Eve had no pre-conception of evil prior to Satan appearing on the earthly scene. Man, being free to choose is not misled by what Satan says. No, man is misled by his own desire to "believe" what Satan says. The final choice is man's own. His effectiveness is always based on man's willingness to follow his "sweet" temptations, and he knows every weakness we have in the flesh. He, along with his demonic spirit beings, working with his human agents here on the earth, has gained total control of worldwide political leadership. Jesus himself said, "He (Satan) is the "God" of this world system of things." The entire world of mankind is under his power and authority. This means he has given his power and authority to the world governments. Yes, the very minute we renounce *Yahweh's* sovereignty, we become one of Satan's earthly agents of evil. We have then accepted his number. Many people are willing to exchange the truth of God's word for the "lie." They would rather worship that which was created, instead of the one who created all things. This is especially true in the scientific community where their real message is the denial of the existence of a creator. Does the pot say to the potter, "You did not mold me?"

Very soon now, "all" will come to know that *Yahweh* really is the grand creator. As he himself declared, "I shall certainly sanctify my great name, which is being profaned by all the nations, and the nations will have to know that "I am *Yahweh*". (Jehovah)

When he confronts the nations of this earth soon, they will know that because of *Yahweh's* indignation, the earth will rock, and not one nation will hold up under his power and denunciation. Yes, we are about to see the fulfillment of his words recorded at Psalms 37:9-11, where we read, "Just a little while longer, and the wicked ones will be no more, the righteous themselves will possess the earth, and they will reside forever upon it." Not forever in heaven, but on earth.

All those who do indeed find the true wisdom of God, and make the wise choice, will survive the coming horrors of the unparalleled great tribulation, and the following war of Armageddon. These protected ones will live into the time of the renewed and restored paradise earth. This will be a paradise of indescribable breathtaking peace and beauty. The entire planet earth, under the direction and assistance of the angels will be restored to its original paradisiacal condition. In many of the prophetic books of the Bible, divine promises are to be found regarding the restoration of the earth. These prophecies show that paradise conditions are to involve people themselves, who by their faith and loyalty to *Yahweh,* would now "sprout" and flourish as trees of righteousness under spiritual prosperity. The eating from the "tree of life," in the paradise of God, is a privilege to be granted to the "him that conquers." This eating of "the tree" means life eternal to those so privileged. These conditions will prevail because the entire earth will be filled with the true knowledge of *Yahweh* as the waters cover the sea.

Even the wild animals of the field will coexist side by side in total peace and harmony with humankind. Even a mere child will be leader over them.

Does this not well describe what mankind is searching for during his short life span here on earth? This hoped for dream of man can never be realized unless we find that true knowledge of God, and use that knowledge in making the right choice. We have the most powerful forces in the universe willing to guide us, the alien spirit world of "*Yahweh.*" The angelic warriors do not make mistakes.

The Bible Book of Revelation is particularly revealing as to the basic reason why mankind has been suffering for so long, and "why" a loving creator has allowed this to continue for so long. Satan, the rebellious angelic being, challenged *Yahweh* by asserting that no human creature would maintain his integrity to him if put to the test of suffering. That challenge had an impact on every human being living on the earth. *Yahweh*, to prove Satan to be the liar he is, has permitted Satan to "roam about in the earth which has resulted in unrestrained wickedness and suffering of man.

Satan has manipulated and arranged his deception in a way, which made it appear that *Yahweh* was responsible for man's hardships and suffering. The question Satan posed to *Yahweh*, was, "mankind will not continue to serve God out of love if his (mans) calamities are severe enough."

How many times have you heard someone "blame" God for some calamity they may have suffered in life? Some will say, "Why did "God" take my only child?" Such misdirected reasoning is never based on true knowledge of God. They are clearly unfamiliar with the issues raised by Satan. How do you view this? The bible writer James said, "When under trial, let no one say, "I am being tried by God," for with evil things God cannot be tried nor does he himself (God) try anyone." Jesus Christ, the only begotten, firstborn Son of God, knew that Satan both exists and is the cause of man's suffering. He also knew that Satan was a troublemaker in the heavenly spirit realm as well. The Bible Book of Revelation focuses specifically on how Satan harassed the spirit creatures so profoundly, that he succeeded in convincing one third of the angelic beings in the heavens to join him in rebelling against

Yahweh, and the angels remaining loyal to him and his spirit son, Michael. This heavenly rebellion caused so much disturbances in the spirit realm that *Yahweh* allowed his son (Jesus) to initiate warfare against Satan's hordes. You can read of this war in heaven at Revelations, Chapter Twelve, verse Seven where it reads, "And war broke out in heaven. Michael and his angels battled with the "dragon" and it's angels. And the dragon and its angels (demons) battled, but it did not prevail, neither was a place found for them any longer in heaven in the 1914CE period according to bible record chronology. Continuing the account, "So down the great dragon was hurled, the original serpent, the one called Satan (and devil), who is misleading the entire inhabited earth, he was hurled "down to the earth," and his angels were hurled down with him." The heavens have now been purged of Satan, his demon angels, and any further influence he might have had on the remaining favored angels in the heavenly spirit realm. They can no longer enter into *Yahweh's* presence again. Continuing, "And I heard a loud voice in heaven say, "Now have come to pass the salvation, and the power, and the kingdom of our God (*Yahweh*), and the authority of his son (Christ), because the accuser of our brothers (the other remaining angels) has been hurled down, who accuses them day and night before our God." This action resulted in the enthronement of Christ Jesus over the entire heavenly spirit realm, never to be smitten by Satan and his followers again. Verse Twelve states, "Woe for earth and for the sea, because the devil has come down to you, having great anger, knowing he has a short period of time." Notice here that Satan is now misleading the earth, and has "great anger" because he knows he is to soon be cast into the abyss where he will remain out of communication with man or angelic beings for a thousand years. Knowing this, he has, since 1914CE, been successful in converting the earth to total evil and wickedness. Since that time, the earth (mankind) has seen all the signs given in Chapter twenty-four of the book of Matthew, including two "world" wars and hundreds of minor wars and armed conflicts. Satan's pressures on the world of mankind in these, the last days of world society as we know it, is indicative of why the time is now for us to make the right choice, and live.

The Bible prophecies show that these events are to proceed and mark the time of the coming "great tribulation" which will envelop the entire world of mankind. This extraordinary time of world distress will also mark God's time of decisive intervention in human, political, and ecclesiastical affairs. He will put an end to the present corrupt wicked world system of arrangement. We are assured however, that some will "survive" this time of tribulation. What, if anything, can we, as people, do while we await this transformation which will bring total relief from human suffering? Yes, wherever we live, whatever we do, or whomever we are, we can give praise to our grand God and creator, and ask him to help us in "making the wise choice.

CHAPTER 10

THE FINAL SHOWDOWN

After more than thirty years of investigation and research into possible earth visitation by alien spacecraft, known as UFO's, I have contacted, and interviewed more than a score of ordinary untrained observers as well as professional witnesses who have been eye-witnesses of these unexpected, unusual craft. Based on the evidence and information collected, it is my conclusion that this extra-ordinary phenomena is real, and cannot be regarded or dismissed as some unexplained mental delusion or fabricated hoax on the part of the observer. One on one interrogation data, in every case studied, revealed that the craft had unusual and uncommon flight performance capabilities. This includes extreme maneuverability in all points of the compass, vertical flight, extreme g-force acceleration, extreme deceleration, blink-out, noiseless and other super-natural characteristics.

Many observers reported that the craft was seen to go from a hovering position near the surface, "dead still," to "out of sight" in the blink of the eye. This sudden and sustained acceleration into "space" would clearly rule out the possibility of a humanly operated craft. No human body could survive such high g-forces under any circumstances. This one fact in itself proves the vehicles are piloted by an entity other than man. This fact also supports my contention that these space chariots are in fact being manned and operated in our atmosphere by spirit angelic beings, which are able to withstand such radical acceleration forces. Also, The craft itself must be structurally capable of withstanding such very high g-force loads.

After questioning "near observer witnesses", most were quick to point out that the near presence of the spaceship had an immediate, or delayed, physiological or psychological result, but nothing disabling or of long duration as far as they could tell. None could recall being abducted or directly communicating with them. However, one hundred percent said they had a strange feeling, or sensation at the time of the encounter.

An example of one near sighting case investigated, involved a mother-daughter incident near Lake Norman, North Carolina.

The mother, a local schoolteacher, and her fourteen-year-old daughter were in the front yard of their home working in the flowerbeds when they spotted a star-like light high in the sky. The time was late evening, near twilight. They paid no special attention to the light until they noticed it had gotten larger and now appeared a reddish orange sunglow color. The object came directly down to about 200 feet above them, overhead. At this point, the mother said the "thing" was a very large bright light. She reported that she knew at this time that it was something unusual. She said the object was so intensely bright, they could not see a shape in the brightness of the light. However, after shielding her eyes with her hand, she could recognize a large triangular shape within the brightness. She said at this time that an even brighter, extremely intense light beam enveloped them. The spacecraft appeared to be 200 to 300 feet long as measured along the forward angles. She reported the light beam gave her and her daughter a tingling sensation similar to a "niacin" blush. The craft remained directly overhead for what seemed to be five-six minutes, until she screamed to her neighbor across the street, at which time the "thing" began to move forward in front of them, very slowly at first. And then zoomed away in a steep ascent and in the blink of an eye it disappeared. By this time it had gotten near darkness. The entire episode was strangely silent. Both observers said they had an eerie feeling during the entire experience. The mother also reported that both her and her daughter had a tingling skin sensation that lasted about three days after the event. The mother also said both of them had a premonition they were going to be "picked up."

This is a condensed version of a typical sighting report of those who have been exposed to a near observation of the spirit controlled "space chariot."

The psychological impact upon the human psyche when exposed to spiritual power is not new. When the prophet, Daniel, received his divine prophetic message from *Yahweh*, the "first cause," through the angel Gabriel, as recorded by Daniel, Chapter 8 and 9. Daniel was so affected by the encounter that he fainted (from fear) and could not even speak until he had been emboldened by the angel Gabriel. Daniel was so terrified by Gabriel's presence that the angel had to touch Daniel's lips before he would even speak. His inner strength was immediately restored by the angel's touch, and he was able to talk. And this happened to a spiritually strong servant of God, who was expecting the angel's visit. The two incidents cited, well demonstrate the super-natural power radiated by the alien spirit being in the near presence of the human creature. It also illustrates the powerful influence the spirit creature has over human consciousness and thought.

Because *Yahweh* created his spirit sons with much higher mental ability than man, they are far superior to man, in all aspects of advanced technological development and application. This was well demonstrated in the nuclear destruction of Sodom and Gomorrah. Their mental, physical, spiritual, and technological powers are only limited by *Yahweh* himself. As *Yahweh's* messengers and ministers, the angelic beings have been able to enjoy many special privileges over the billions of years they have existed in the universe. They are technologically capable of traveling to the most remote part of the vast universe in an instant. They are not limited by time, distance, and speed, unlike man. Angels, by *Yahweh's* will, "are time, distance, and speed". Time, distance and speed (or velocity), is neutralized completely by the eternalist, *Yahweh*, (Jehovah, and Elo-him).

From earth's beginning, the angels have been an active part of the progressive development of the humankind creation. Over the past six thousand years they have been visiting the planet earth hundreds of times. Some included visits to Abraham, Daniel, Jacob, Moses, Noah, Joshua, Isaiah, Peter, Paul, John, Ezekiel, and many others. Their visits contributed much to the development of earth, and in the writing of the "Holy Bible," which are in fact *Yahweh's* instructions to mankind. It was an angel "flying" in mid-heaven (earth's atmosphere, where birds fly), that delivered to man the "good news of Gods kingdom", His word tells us, regarding the "work" of angels, "Are they not all spirit beings for public service, sent forth to watch over those who are going to inherit "eternal life"?"

God's true followers on the earth are assured by Yahweh, that they are under the full protection of his army of angelic spirit beings. He has commanded his angels to "guard you in all your ways." Yes, his heavenly host, the alien spirit beings are "camping out around all who fear him," ready to rescue you from the evil one, Satan.

It is utterly impossible for mankind to grasp the reality of *Yahweh's* wisdom and power, because his will and purposes are so much higher than mans thinking level. He made this clear when he said, "My thoughts are higher than your thoughts, and my ways higher than your ways." Unless a person exercises "faith" in the truth of Gods son, one can never discern the full depth or the magnitude of his written word, nor can one understand how his stated purposes are in total harmony with what he foretold, and had recorded thousands of years ago by his inspired ones here on earth. Paul said, "Many will forever be learning, but will never come to gain the accurate knowledge of the truth". (The reality of *Yahweh's* unmatched superiority).

If we would direct our thinking abilities into a proper application of learning, we will come to know the very wisdom of *Yahweh*. This would serve as an impenetrable shield against our being ensnared by the death-dealing enticements of Satan's world.

The wrong kind of knowledge, worldly knowledge, can only bring us vexation, pain, suffering, and death, in a very short life span. *Yahweh* warned us, "Knowledge gained by the devotion to many books (worldly books), is wearisome to the flesh, and does not preserve the life of the soul (The life of the body)."

What can be learned from these incredible examples given to mankind, by *Yahweh,* in his recorded word, the Bible? We see that he has a predestined purpose for sending his spirit sons (warriors) to the earth at this time, in this generation. He has clearly made it known to us that he is sending his angels here to fulfill his predetermined will and purpose for the betterment of mankind. This is provable reality. And we will see an ever-increasing number of visitors as we draw closer to the end of this world system of things. Very soon now, these angelic warriors will reveal their true identity, and their "mission" here, to man's shocking surprise.

It is also clear, that the time remaining for us to gain that life-saving, true knowledge, offered to us as a reward for our loyalty and integrity is very short. The decision is ours alone. The final result being "Search, and you will find. Knock and it will be opened to you. Ask, and it will be given to you," said the angel "Michael," the son of the true God, *Yahweh*.

More than three thousand years ago, a large self-propelled space ship (God's chariot) was seen flying over southwest Asia. This was long before man was able to fly. It was not an invention of man, nor was it mere imagination. To the observer, it was an awe-inspiring experience. Nothing like it had ever been seen by man up until that time and recorded a very detailed description of the flying vehicle. Its design far exceeded mans technology even today. Ezekiel, a prophet of *Yahweh*, was the observer of Gods' chariots and recorded the event for all mankind.

Ezekiel the reporter, and the observer, even tells us where he observed this strange spacecraft. He said he was standing near one of the canals near the Euphrates River in Babylon. What he saw was no illusion, he said, "And I began to see, and look! Coming from the north, a great cloud mass of quivering fire, and it had a brightness all around, and in the midst of the brightness, was something like electrum, out of the midst of the fire. And out of the midst of it (the craft), came four living creatures (Angelic aliens). And they had a likeness of earthling man." The angelic cherubs were capable of directing their movement to the four cardinal points instantly without turning at the speed of "lightning," denoting unlimited speed. They are not constrained, or bound by speed, time or distances.

As for the spacecraft itself, Ezekiel said, "They glowed like chrysolite (were highly polished), and the appearance of their structure (their design), was like a wheel in a wheel (circular in form with an inner circular structure (typical UFO design seen today). As for the outside circular rim, it was full of eyes." (This denotes lights around the perimeter of the "wheel." The craft were so large as to cause "fearfulness." Ezekiel reports that the spacecraft would not "be lifted" up unless the cherubs were "in the wheels." The spaceships had no external, "wings," but could rise into space with ease. There is no doubt that they are powered by an anti-gravitational generating unit. This is not as complicated as it might appear. It could be made possible by designing an Electro-static generator, which would produce a static particle, which would be of the same polarity (plus or minus) that the earth's magnetic field carries. Remember, like charges "repel" while opposite charges "attract" (and will discharge or neutralize). When our spacecraft has been statically charged to equal or exceed the field strength of earth's Electro-magnetic force (measured in gauss), the craft would be repelled from the *EARTH* into space. The particle +, -, would be collected on the outer skin (surface) of the vehicle. Remember too, that earth's, and all other celestial bodies' magnetic field lines are concentrated at the north and south poles. A vehicle so powered would have unconventional flight capabilities.

In any event, *Yahweh's* cherubs have been provided with the most advanced technological propulsion system possible.

Yahweh, who created and established the visible universe, has also created and organized all the heavenly spirit angels into one vast universal family over which he reigns as sovereign ruler. Each one of the thousands of angels has been assigned his own place in the celestial arrangement, and each one has "his own" God-given duties to perform. They serve as his divine commissioned ones always ready to respond to *Yahweh's* will and purpose. Some are ranked as archangel, cherubim, seraphim, and angels. The spirit creatures and the "space chariots" as seen by Ezekiel at the river Chebar is a mere divine manifestation of an enormous, angelic, chariot equipped organization, extending to all areas of the created universe.

All this is ample evidence that *Yahweh* has the unlimited power, organization, and will to reclaim planet earth from Satan's control and domination. This conflict is truly the final showdown.

CHAPTER 11

WHY HERE?

The earth is the fifth planet of the solar system, and the third in the order of position from the sun. The planet earth rotates on its axis, which results in day and night. The tilt of earth's axis is 23 degrees 27 minutes and the earth is held steadily in this position by gyroscopic force resulting from the effects of rotation. Our earth's atmosphere is ideally balanced for human breathing, and is composed of nitrogen, oxygen, water vapor, dust, and a number of other rare gases, which extends more than six hundred miles above the earth's surface. Beyond this is outer space (the heavens). The planet earth is the special place *Yahweh* prepared for man's occupancy, and provides everything in abundance for mankind's survival. Its original purpose was to be a paradise for all living creatures on the earth.

Its creation was simply stated at Genesis 1:1 where we read, "In the beginning God created the heavens and the earth." Just how long the creation process continued, was not revealed to man. Current scientific studies of the earth's rock masses indicate the age of planet earth to be at least four thousand million years old. This estimate is no more than an "educated" guess. However, We do know by the Biblical record that "man himself" was created by *Yahweh* and his son a little over six thousand years ago. (Which ended in 1974-5 CE). When Paul spoke to the Athenians regarding man's creation, he said, "He (God) made out of one man every nation of men, to dwell upon the surface of the earth." Hence, no record of ancient men before 4027 BCE exist (The date of Adam's creation). Based on Biblical chronology, 6000 years of mans history ended in 1975 CE since the Bible record traces mans history back to Adam's creation (see Luke 3:24-38). There could not have been an ape-like prehistoric man. This reality explodes the evolutionary hypothesis as mere human speculation, with no basis of facts. Fossil findings in the earth do not support a link between ancient animal species and a prehistoric man. How could animals have survived the catastrophic flood of Noah's day without divine guidance? The flood event is true and provable. Where is proof of prehistoric animals becoming prehistoric man through a process of blind evolution?

Bible scholars generally agree that all the real evidence, which taken from Genesis archaeology, the traditions of men all point to the Mesopotamian plain as the oldest home of mankind, the cradle of civilization. Remember too, man is the image of God in that he was "created" with moral qualities like those of God, namely "love and justice," with wisdom far above animals. No other living creature on the earth has, or ever will, possess these divine qualities. In reference to "time," the creative days did not include the earth itself, but was the time involved in preparing and arranging it for human habitation. *Yahweh* does not reveal to us whether he created life on other heavenly bodies. Jesus did reveal to us that, "In my fathers house (the heavens), there are many abodes (A place of residence or a dwelling place)." Jesus was really telling us that there are many places to reside or dwell in the heavens, or "outer space." We do know that millions of spirit beings reside out there in the universe.

Like all of his wonderful works, *Yahweh* did not create this earth without a purpose. He brought the earth into existence for "his pleasure," and he installed the earth to remain until time indefinite. When he placed man on the earth, it was perfect, without flaw. It was his will that man live upon the earth in total happiness forever under his divine leadership.

He also told man that he was "hanging the earth upon nothing." This truth was verified when man was allowed to travel out into "near" space. Man was then able to look back at the earth "hanging on nothing." He also informed us, in his word, that the earth was circular in form. This too was verified from space. For hundreds of years, man believed the earth was flat. *Yahweh* had told man thousands of years ago, that the earth was suspended on nothing (space) and was globular in form. Man just refuses to believe, does he not? He will accept the truth only when it is forced upon him, having no other choice.

Yahweh had a definite purpose for creating the earth. He said, "I did not create the earth for nothing. I, even I created the earth to be inhabited, yes, even to time indefinite." Having placed the newly made creature called man on earth, it was *Yahweh's* purpose that man remain loyal to him, and enjoy a life of eternity in his paradise home. Although the physical creation of the earth is a divine accomplishment in itself, the real meaning of human existence is clearly of greater importance. Each one of us should ask himself the question, is man the center of meaning? The wise man of old, Elihu observed, "Because of the multitude of oppressions, I keep calling for aid; I keep crying for help…And yet no one has said, where is God, my grand maker?" His words underscore the fact, that we humans are not the true center of meaning. Our grand creator is that center, and our existence is totally dependant on him.

Even if the scientific world were to continue to study God's creative works for the next one million years, man would be no closer to understanding the full scope of the universal mechanics of creation. Although man himself is viewed as a special part of the creation, he needs *Yahweh's* mercy and forgiveness to regain his original righteous relationship with him. Sin placed man out of harmony with *Yahweh*. It thereby isolated his relations with

God, and the greater part of the divine creation. It resulted in consequences of alienation from the spirit realm, and death to all men, and therefore death continues to rule as king over mankind down to this day. Even over those who had not sinned after the likeness of the transgression by Adam. The evidence in all reality, shows that the passing down to all succeeding generations, sin, which was due to the "law of heredity." And so, we can now truly recognize the life-saving value of the sacrificial death of the angelic son, Christ Jesus, or "Michael," his heavenly name. Since Christ Jesus was sinless in his life on earth, he was able to offset our condemnation of death, by his willingness to die in behalf of all mankind. The sacrificial arrangement by *Yahweh* was designed to nullify the untold damage and suffering mankind has been subjected to because of Satan's rebellion and his desire to be like his own God, *Yahweh*. Under Satan's curse of sin, mankind had no chance of returning to life in a promised resurrection after the "first death." God's word tells us, "For he who has died, has been acquitted (of) from his sins." True to his word, "The wages sin pays is death." Because of, and through the ransom paid, *Yahweh*, by his loving provision in sacrificing his firstborn son, Christ Jesus, by this grand arrangement, we, upon death, are not only forgiven for our sins, but we repay *Yahweh* our part of the ransom he paid in the sacrificial death of his son. (Read John 3:16; "The first death" releases us from our sinful state of Satan's slavery.)

Jesus assured his followers, "It is the spirit that is life giving; the flesh is of no use at all. These sayings that I speak to you, are "spirit and life." Jesus was showing that the spirit of God is eternal. While under sin, flesh is but temporary. If we as men, really appreciate true knowledge and wisdom, we should strive with our very being to Gain the Holy Spirit of *Yahweh* (His motivating force).

Most people do not know what God's "spirit" really is. The word "spirit" is translated from the Hebrew word, "ru'ahh", or the Greek word "pneu'ma". Both have the meaning of, "an active force," a "remote force," "breath of life." In his design of man, *Yahweh* equipped man with spirit receptors, which would enable him to activate man's inner psyche which serves to incite all parts of the human organism to the demands of his will. His spirit force is very similar For example, to a remote radio controlled space satellite being manipulated and controlled by man, even though the space satellite has been landed on a distant planet, all being accomplished by radio "wave force" from a transmitting source a long distance away. This principal is exactly the way God's spirit acts upon man. God sends his spirit to those whom he chooses on earth for "his" purpose, to be manipulated or inspired to do his will. The "spirit" of God, can accomplish the following when acting upon; It teaches, it bears witness, it gives evidence, it acts to guide one, it acts as a helper, and it can motivate man to do anything *Yahweh* wills.

The spirit of God then, is not a "third person" in the Godhead, as taught in most religions today, but is in fact, the all-powerful force of God. He used this force not only to create all things, but also to "maintain" all things. Even inanimate, physical things respond to his powerful spirit force. We humans are especially programmed to respond to the direction (control) of his spirit, if we are found in His favor.

His glorified angelic spirit son, Christ Jesus, said, "He who conquers the world by "spirit", I will give eternal life from the tree, which is in the paradise of God." This "tree" is the symbolic provision for sustained life on the "new earth."

Yahweh, when he placed man in the Garden of Eden, purposed that man should enjoy a never-ending existence. His divine purpose however was temporarily interrupted by "Satan," the rebellious spirit being. Although *Yahweh*'s purpose for man has never changed, he has allowed man and Satan to continue living while he laid the legal case in the heavens and on earth to expose Satan as the evil one, and eventually destroy him, and his demonic and earthly followers.

This shows the current state of the world conditions in this, the "last days" before the soon to come restoration of mankind back to his original state. The apostle Paul wrote under spirit inspiration, "The spirit says, in the last days of this worlds system, some will deny the truth of God's word and will be led to believe instead, misleading teachings of demons (Cast out angelic rebellious beings)." This is to cause critical times, hard to deal with. The wickedness and evil of man is brought about primarily through the influence of Satan, his hoards, and false apostate teachings by corrupt religious leaders, in the world of Christendom. This world system of false religion is identified in the Bible as "Babylon the great" which commits fornication (supports the political. Elements) with the political leaders of the nations. *Yahweh* warns us, "Get out of her my people, unless you want to share with her in her sins, and unless you want to share in part of her plagues, for the one judging her is strong, and her (false religion) destruction comes in one hour (quickly)."

Michael said, "I do not come to the earth to cause peace upon the earth, I come to put not peace, but a sword. My coming will cause division…However, everyone who comes to know *Yahweh* (receives his spirit) will be spared. How will they call upon him to whom they have not put faith? How in turn will they put faith in him if they do not know him? How in turn will they know him if someone does not tell him my name, unless I have sent him forth to tell you? And so faith follows the thing

> heard, in turn, the thing heard (the good news)
> is through the word about Christ."

Our grand creator, *Yahweh*, Jehovah, or Elohim will soon demonstrate his power and sovereignty toward all his creature beings, both human and spiritual. He is even now using his loyal angelic beings to prepare the earth and the heavens for his coming kingdom, a government that will never be replaced. The establishment of this universal kingdom will require an enormous effort to "bring all things" back to a proper divine condition, just prior to the utopian transformation. A transformation that will, result in total annihilation of more than two hundred million fighting men from all the nations who resist Michael and his heavenly angelic warriors.

The apostle John, in his written account at Revelation, Chapter fourteen, gave clear details of this final refining work to be done on the planet earth by the angels of *Yahweh*. Here John said, "And I saw, and look! A white cloud (war chariot), and upon it someone was sitting like the son of man (the spirit being, Michael) having a golden crown on his head (He had been crowned as "king" over Gods heavenly kingdom). And he had a sharp sickle (he had been given all the authority to separate the wheat-like people of earth from the weed-like people on earth) in his hand."

"And another angel emerged from the temple (Gods sanctuary), crying loudly, saying to the one sitting on the cloud, "Reap with your sickle, for the hour (time of the end) has come to reap (Separation work), for the harvest of the "earth" is thoroughly ripe. (All those who are going to accept the "truth" have now done so.) And the one sitting upon the "cloud" did just so and reaped." (Identified all who were worthy) the weed-like ones sown by Satan were burned, or "destroyed."

The details here revealed to us by the spirit inspired apostle, is but a mere glimpse into the overall divine work being carried out earthwide by the angelic warriors of *Yahweh*, the most high God of the universe. This clearly answers the question, "Why here?"

CHAPTER 12

THE SECOND DEATH

Most people who profess to be Christians, believe that their fate of either life in heaven, or burning in torment in "hell", will be determined by the way they live their "life". This is a reasonable expectation, since most "Christian" teachings direct their doctrinal beliefs toward this central concept. This belief has been branded in the minds of millions of "sincere" sheep-like followers all around the earth. However, is this concept of "heaven or hell" really in agreement with God's word, the Bible? Well, let's examine his record closely and see.

First, we should gain a full and accurate understanding of what the Bible shows hell to be, and just who, and why, people may go (in) to hell. (You can research this for yourself). The English word "hell" was translated from the original Hebrew word, "Scheol," and the Greek word, "Hades." In every instance where Scheol or Hades was used in the word of God, both Hebrew and Greek, the meaning was always intended to mean the common grave of mankind. (All seventy-six occurrences.) Since "hell", "Scheol", and "Hades" are synonymous, this would prove that Scheol, Hades, and hell do in fact have the same meaning; "The common grave of mankind", not the erroneous teaching of "a place of eternal torment." How many people over the past centuries have been misled by this satanic lie? Even Jesus himself was in "hell," (the grave), for three days (Read Acts, 2:31, Ps. 16:10). Does this mean then that those who go to "hell" will never get out? No, again the bible shows that. (Rev. 20:13-14.) "The sea gave up the dead which were in it, and death and hell delivered up the dead which were in them." Here we read the "truth." Note that hell delivers up the dead. Now, does logic not indicate that those delivered up from hell, to be judged, were not in hell in torment? Simple reasoning will expose the lie of false teachers. To the uninformed, even "false" knowledge can be convincing if presented in a persuasive manner. A wise Bible writer recorded, "Make sure of all things." Take no "man's" teachings for the truth, but search the scriptures to see that all these things are true." Clearly there is only one source of truth, *Yahweh's* recorded word. We should be especially cautious when it comes to religious, scientific, or political wisdom, knowledge, or understanding, as all such reasoning is based on man's suppositional intellect, and should never be accepted or acknowledged without suspicion or skepticism. The apostle Paul, in his letter to Timothy said, "In the last days, many will turn away from the faith (truth) and listen to misleading inspired utterances and teachings of demons." For there will be a period of time when some will not want to hear healthful teachings, but will, because of their own evil desires, select teachers (ministers) among themselves to have "their ears tickled", and they will refuse to listen to the "truth," and will only accept "false knowledge" as the truth. But you, though, keep your senses in all things. Since the dead are in the grave, as shown in the book of Revelation, this means that "hell" (the grave) will give up those dead ones in the Resurrection period. A-na'sta-sis, a Greek word translated, Resurrection in English, literally means "to stand up again," or "to rise up from death." This provision of a Resurrection is a loving expression of *Yahweh's* undeserved kindness, and is a clear indication of his wisdom and divine power.

This promised resurrection of the dead is not a re-uniting of an immortal soul with the fleshly body. The Bible does not even imply this to be so. That idea comes from Greek philosophy, and has no basis at all from Bible truth regarding a human soul (being). This misleading Christian concept of an immortal soul being merged with the fleshly body at the time of conception, is a clear example of Christian philosophy. Which has now evolved into an important doctrine of teaching of the immortality concept, where death is nothing more than a transition from one life to another life. The main problem with the immortality teaching is that it is a concept developed by man, and is not taught or even mentioned in the Bible record. When the Bible record truth is combined with philosophical precepts of men, you no longer have the truth but doctrines, customs, and teachings of demons. The truth of God's word is deliberately diluted by such concepts. The uninformed is easily misled, and will follow blindly. Jesus, God's son, said, "Where the blind are leading the blind (unlearned), both the leader (teacher) and the one being led will fall into the pit." Many people, in this world's religions today, are indeed being led to the pit (death) unknowingly. They are quick to defend their "religion," because their "ministers" have convinced them. When in fact, all he has convinced them of is just how "blind" they really are. Some will surely be offended for sure, but some will be awakened to the truth, and perhaps live.

The only way we can gain *Yahweh's* spirit and favor is by his undeserved mercy. This mercy is shown if we are fully committed to accept his Godly direction without reservation or qualification. This means we are willing to "repent" for failures of life, and demonstrate our desire to "change" our willful practice of sin, and ask for his forgiveness with a pure and searching heart. Our sincere repentance marks a total change in our view of this evil system of the world. Repentance must involve one's whole life course that is in opposition to God's will and purpose. This means we must change our thinking with regard to past, or intended future actions, conduct, desires, self-control practices, wrong tendencies, inclinations, and attitude. Even though we may take a false step due to our fleshly weakness, we are assured by *Yahweh,* that he is aware of our shortcomings and will forgive our errors resulting from inherent fleshly imperfections, ignorance, or forgetfulness. We must, however, strive to acknowledge our failures in living up to his righteous standards, and ask his forgiveness and mercy repeatedly.

A Godly consciousness is one of the most profound mysteries of the human existence. The human "mind" is an elusive entity wherein knowledge, awareness, thoughts, images, sounds, memories, feelings, and hundreds of bits of informational data is fed into or through, in a constant flow. Our most innate personality is a product of the large quantity of data flowing into the mind. That is why Jesus said, "You must love *Yahweh*, your God, with your whole mind." Do not keep your "mind" on the things of the earth or the flesh, no, keep your mind on the things above." As we begin to focus on the things above, through God's word, our conscience is no longer being flooded and diluted by the false counterfeit knowledge of Satan's world. We will immediately see a change in our desires, attitude, and values. Our entire "personality" then changes. Again Jesus said, "You are now stripping away the "old personality" which conforms to your former course of conduct, which is being corrupted by "deceptive desires." But now you are "being made new in the force (spirit) actuating your mind." You now have a new mind and a new personality. You will now receive *Yahweh's* powerful spirit because you have, by you own choice, become one of his "special servants," and are now a chosen one in his kingdom. Even his angelic beings now recognize you as "their friend." It is only then, that you, as a human, will begin to comprehend your true reason of existence.

The apostle Paul recorded at Romans, 8:12-14, "So then, brothers, we are under obligation. Not to the flesh, to live in accord with the flesh, for if you live in accord with the flesh, you are sure to die, but if you put the practices of the body to death by the spirit, you will live. For all who are led by God's spirit are God's sons."

In the Bible Book of Revelation, John gave us a clear insight into the specific meaning of the "second death." The symbolic reference to the "lake of fire" has no reference to "hell" as translated from the original Hebrew and Greek words "Scheol," and "Hades." The "lake of fire" into which death, Hades (hell), the symbolic "wild beast," the "false prophet," and the angelic being "Satan," with his demonic and human followers are cast, is shown to represent the "second death." This second death is separate and distinct from the inherited "first death." It is clearly evident that all those who are destroyed in the "lake of fire," which is the second death, are in fact annihilated, and no longer exist in any state or form, and so, to be cast into this symbolic lake of fire means total everlasting destruction.

Those who gain the "crown of life" through the "first resurrection," are free from the authority of the "second death," and receive immortality in the kingdom of God. (Rev. 20:6.) While those who are given life again in the "second resurrection," will receive life on the earth for one thousand years, after which "Satan" will be released from his prison. He will again be allowed to test (tempt) all of those who were brought back to life in the "second death." Read Rev. 20:7-9.

Since no other name has been given under heaven by which man may be saved, all those who are searching for the "real life" must acknowledge and follow the "mighty archangel", Michael. Anyone who wishes to become a "friend" of Jesus Christ must follow "his way" and safeguard his own heart. Jesus counseled, "Let him that thinks he is standing firm beware that he does not fall. More than all else that is to be guarded, safeguard your heart, for out of it are the sources of life."

At the beginning of man's existence, it was never purposed by *Yahweh*, man's creator, that man die. Since death was never intended to be a "natural" part of the human experience, no resurrection arrangement was necessary. Consequently, death was introduced into the human family as a consequence of man's willful transgression. This act of transgression by Adam has subjected "all men" to death. It is through the "sacrificial death" of *Yahweh's* first angelic son, Christ Jesus, that mankind was taken out from under the authority of death eternal, through the hope of resurrection. The only dead human creature who does not receive a resurrection, are those who were directly executed by *Yahweh's* pronouncement.

Those awaiting resurrection in their grave or place of burial are totally oblivious to the passing of time. From the time of death (cessation of life), until resurrection is, as "the blink of an eye", even if the "time" involved is in the "thousands" of years, to the dead ones, it is as if the transition is immediate. Only the living are aware of time passage, "But as for the dead, they are conscious of nothing at all," and thus the fleshly body of the dead returns to the earthly elements from which it was composed. The life force (spirit of *Yahweh*) contained in the body's "blood," has now been reclaimed by *Yahweh* who gave it. This is proven to be true as shown at Ecclesiastes 12:7. "Then, the dust (body) returns to the earth just as it happened to be (before conception "to life"), and the spirit (not soul) itself returns to the true God (*Yahweh*) who gave it." Psalms 146:4 says, "His spirit (life force) goes out, he goes back to his ground, in that day his thoughts do perish." Note his "soul" does not live on into a state of immortality in some ethereal location in the heavens, but has now "died" his "first death." Through the prophet Isaiah, God promises mankind that at his "appointed time", death itself would be swallowed up forever. His promise includes both the first and second death. Clearly the abolition of death demonstrates that which causes death, "sin," is also abolished from obedient mankind. Therefore, death, which was brought upon the human creation by Adam's transgression, "will be no more." All those who gain *Yahweh's* favor after the one thousand-year reign of Christ and "Satan's final test," will receive immortality in the glorious new earth (read Revelation 21:1-8), and will not suffer the "second death."

CHAPTER 13

STRANGE VISITORS

An unexpected encounter with an unknown extra-terrestrial spacecraft from another realm of existence is no doubt one of the most shocking experiences a man could ever be confronted with. A face-to-face encounter with supernatural spirit beings can be an extremely intense, if not perplexing reality. Thousands of sincere, ordinary people the world over are reporting such encounters every year, while many more encounters go unreported.

The following "near" encounters are but a few condensed examples of UFO sightings, I personally, have investigated over the past four years.

Extra-Terrestrial Encounter Number 1

A retired farmer, eighty-six years old, who lives on Highway 70 west, out of Statesville, North Carolina, reported to me, that late one evening, he and his neighborhood friend were standing in his back yard talking, when they noticed a bright light behind his chicken house, which lit up the woods between them and the farm field beyond. At first he thought the light was his son plowing with the tractor, as he had been plowing in the field all day. He thought at first, his son was trying to finish up before dark. However, as full darkness fell and the light got brighter and brighter, they began to have doubts about the son still plowing as they could not hear the tractor's engine, and the light was now too bright to be tractor lights. After talking about how bright the light was (it now lit up the chicken house and back yard); they decided to walk down beyond the chicken house and see what the bright light was. As they passed beyond the chicken house, they were surprised and amazed to see a large triangular shaped "flying machine" sitting about twenty to thirty feet off the ground over the plowed field. There was no noise (racket) and the light given off by the "thing" was so bright you could hardly see the "machine." As soon as they passed to the other side of the chicken house, about fifty yards from where the "machine" was hovering, the triangle began to rise straight up, very slowly at first, and after only a few seconds, the large triangle (estimated to be 200 feet long), began to "speed up", and in a few more seconds it was out of sight, going straight up. No noise was heard at any time, and both had "a funny feeling" after the event was over. They did not report the incident to local law officials.

Extra-Terrestrial Encounter Number 2

A student pilot was standing on the aircraft-parking ramp at the Statesville, North Carolina airport, watching local aircraft shooting landings, through his 7x50 binoculars. As he followed an aircraft (Cessna 172) on take-off climb-out, he was surprised to see a UFO enter his field of vision from left to right. As he then trained his binoculars on the object, it passed over the aircraft at a high rate of speed. The object was globe shaped, resembled a Volkswagen, and appeared black in color. It appeared to be just beyond the aircraft, which had just cleared the end of the runway, but had not made the first left out of pattern turn. The UFO was at least as large as the aircraft and traversed from left to right (out of sight) in approximately fifteen seconds. Only aircraft engine noise was heard. The observer of this UFO event is a field investigator with the

MUFON Network (Mutual UFO Network),
and is a trained observer.

Extra-Terrestrial Encounter Number 3

Two deer hunters were hunting in Northern
Iredell County, North Carolina, and were
returning to their pick-up truck just before dark
one evening. As they exited the woods into an
open cleared field, an extremely bright, white
light came from directly overhead and
suddenly enveloped them. Within the
brightness of the light they could see a large
triangular shaped spacecraft. The sudden
appearance of the craft so frightened them, that
they were absolutely terrified. Although both
men were carrying high-powered rifles, they
were so overwhelmed by the unexpected,
unbelievable event that they said they didn't
even think about shooting the strange craft.
Their main concern was to get to their truck
and get out of there. As they ran for the truck,
they UFO very slowly rose in height and shot
away to the northeast. It was gone in seconds,
even before they reached their truck. This UFO
sighting report was sent to the local newspaper
but was never printed. They said the large
spacecraft was a "dark" color in the light, and

made no noise at any time. It seemed to be no more than 200 feet above them and was very large in size. On their way home, they discussed how something that large could float in the air and move so fast without making any noise.

Extra-Terrestrial Encounter Number 4

A student at the University of South Alabama, Mobile, Alabama had been to lunch and was returning to class. While she walked across the golf course to her class location from the on-campus mess hall, she noticed a circular shaped UFO approaching her at about a forty-five degree angle in front of her. She reported that the flying craft stopped in front of her about 200 feet high and moved in a back and forth rocking motion. The Frisbee shaped UFO remained in this position for five to six minutes and then moved away along the tree line, still rocking back and forth until it disappeared from sight behind some large oak trees in the distance. She said the entire episode was observed by a number of students, and the event made her extremely nervous, and had left her with a feeling of mental exposure. As she returned to the classroom her professor noticed her unusual conduct and asked her to speak with him after class. After class the professor asked her what had upset her, so she related the entire incident to him, and he revealed to her that others had witnessed the same event. A written report with a detailed drawing of the UFO's shape, size etc., was filed with the Foreign Technology Division, at Eglin AFB, Fort Walton Beach, Florida. The graphics branch at Eglin prepared the drawings of the craft, for the report filed.

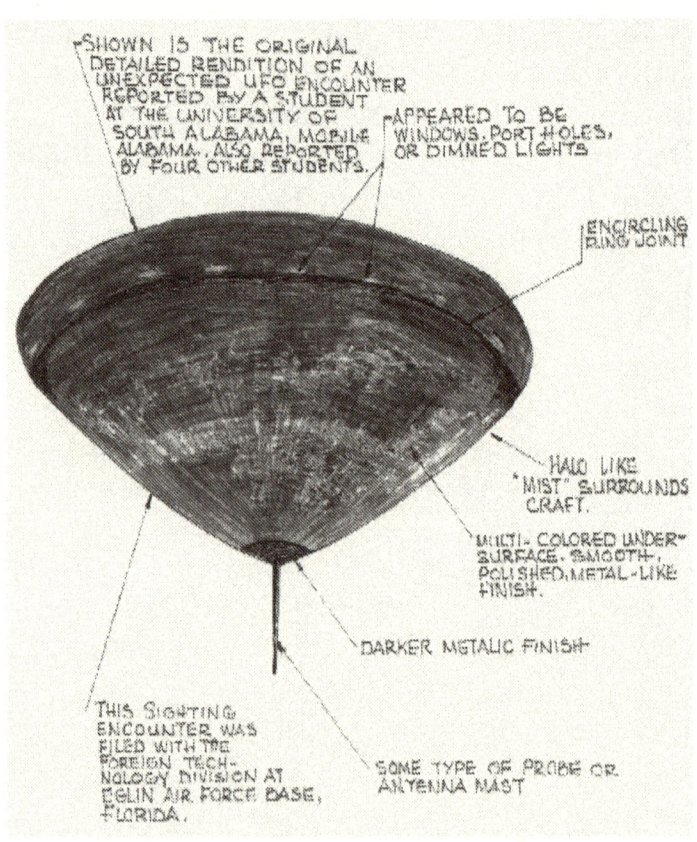

This is an artistic rendering of an unexplained_celestial spacecraft seen near the McGuire Nuclear Power Station on Lake Norman, just north of Charlotte, North Carolina, by two employees on their way to work about daybreak. They reported the craft as moving very smoothly through the air in total silence as it passed directly over their car at about 100-150 feet.

This is an artist portrayal of a widely seen and reported triangular shaped celestial spacecraft. This painting was based upon observer drawings, sketches, and descriptive details given by the witnesses, and is typical of the triangular design seen worldwide. This reported sighting was made near Destin, Florida and is very similar to a craft seen at Lake Norman, North Carolina.

Clearly, extra-terrestrials are visiting our planet earth, as the condensed summation of the four extra-terrestrial encounters reported here would prove. All of the remarkable narratives given of their experiences are real, and show just how common this alien implication really is. Although the people involved in the reports cited here have asked that their names be withheld to assure confidentiality, the reality of the events are not diminished at all. Even though the circumstances surrounding the various encounters were distinctly different, there was a thread of common linkage in all the sightings in a number of ways. This would indicate that the many spacecraft being seen are from the same alien source.

If one truly believes the bible to be the word of the true God, *Yahweh*, then we must also believe that extra-terrestrial visitors are also of and from the same God, *Yahweh*. If the visiting beings are not proven by the Bible record, then the Bible record itself becomes suspect to its reliability. This imagined controversy is resolved by *Yahweh* himself. Since he said, "My word is the truth." There can be no controversy as all things are judged by the truth, that is, all things are judged according to the way things really are, not on the basis of argument or outward appearance. The creative works testify to the fact that God exists. Even though those people who have come to "know God's reality" have chosen to suppress this truth, have been willing to exchange this truth for the lie.

Jesus said, "I have many things yet to say to you, but you are not able to understand them at "the present time." We will never understand the full impact of the UFO (angelic visitation) reality until we have gained the wisdom of *Yahweh,* as revealed through his word and spirit.

CHAPTER 14

DOCTRINES OF DEMONS

Can there be a connection between the UFO phenomena and the demonic realm? To answer this question truthfully, we must first have an understanding of the spirit world, and the spirit beings' influence on human affairs. First of all, demons are invisible, wicked spirit angels or fallen angels, who have, by their own choice, chosen to oppose the purposes and will of the true God, *Yahweh*. They were not created in the beginning as demons, but were angelic sons of God. Having been created as free agents, they themselves chose to forsake their original position in the heavens. They chose to follow "Satan," the first angel to rebel against *Yahweh*. As a consequence of their rebellion, the demons have been severely limited in their sphere of spiritual operations.

Demon angels are now retrained in a condition of dense spiritual darkness and are now incapable of materializing into "physical" state. Their existence is now confined to the invisible spirit state and cannot be "seen" by the fleshly human being. This imposed limitation on their spirit power, however, does not mean they are weaklings. No, even in their limited spiritual state they possess great power and influence over the human mind and lives. They even have the power to enter into and take full possession of the human temple, the body, as well as using "inanimate" objects to influence the human mind. Remember, the whole purpose of demonic activity is to win over the mind of man, and turn him against *Yahweh*, and destroy "pure worship" of the God of creation. That is why *Yahweh* has forbidden demonism and demonistic teachings in any form or practice on earth.

Demonic influence in human affairs today is clearly seen. The nations of the earth are making many sacrifices to the demon hordes in almost every expression of mankind today. Why has the world of mankind been so overwhelmed by demonic force and influence in just the past few years? The answer is given to us in Revelation, Chapter 12:7-9,12, where we read, "And war broke out in heaven: Michael (Jesus) and his angels battled with the dragon, and the dragon and (his) angels battled. But it (he) did not prevail, neither was a place found for them any longer in heaven, so down (to the earth) the great dragon was hurled, the original serpent, the one called devil and Satan, who is misleading "the entire inhabited earth", was "hurled down to the earth", and his angels (demons) were hurled down with him. On this account, be glad you heavens, and you who reside in them! "Woe for the earth, and for the sea, because the devil "has come down to you, having great anger, knowing he has a short period of time."

At the end of the year 1914 CE, Bible chronology verifies that date as the time this war in heaven took place (war on earth also), and Satan and his demon angels were thrown out of the heavens, and down to our earth. Satan, having great anger, not only because he lost that battle, but also because he knows his days are now numbered, before him and his evil followers will be eternally destroyed in the "lake of fire." He is going to do everything in his power to mislead and destroy as many on earth as he can. Hence, ever increasing wickedness on earth since 1914 CE, which was the end of the "appointed times of the nations," or the "end of the gentile times." Read Luke, 21:24-28. The gentile times ran for 2520 years beginning in 607 BCE running through 1914 CE. This was the time *Yahweh* allowed man to rule himself without his intervention. It was to prove to mankind that man could not even direct his own steps. It was never God's will for man to rule himself as he knew greed would bring hardship and suffering to mankind, with self-destruction his final destiny. Read Matthew 24:21-22.

The demons will be instrumental in leading the nations en-masse to the battle of Armageddon against Michael and his heavenly angelic warriors. True Christians must therefore put up a hard fight against these unseen demonic spirits. *Yahweh* warned us, "You cannot eat of *Yahweh's* table and at the same time feed from the table of the demons."

Demon possession is the captive control of, and complete influence over a person by an invisible wicked spirit being. Demonized persons who are afflicted may appear dumb, blind, possess superhuman abilities, foresee future events, and other unusual attributes. Anyone can fall victim to demonic possession if we allow our spiritual suit of armor to wear thin. The agony of an afflicted person can be greatly compounded when a number of demon spirits gain access to the same human being. When a demon spirit is expelled, the person returns to a normal state of mind. Demonic possession will totally destroy a person's spiritual conscience by replacing our "inner voice" warning system. Some of the greatest miracles Jesus performed included setting possessed persons free from demonic captivity. The "son of the living God" has total authority over all the demon spirits including the ruler of the demons, Beelzebub (Satan).

Since the demonic angels no longer have the supernatural ability to transform or materialize into the physical or visual (visible) realm, they cannot be seen by human eyes. This single fact establishes convincing evidence that the UFO (celestial chariot) is not a vehicle of the demonic spirits, but are in fact *Yahweh's* holy angelic spirit beings.

The invisible demon spirits are behind many of the false doctrines being taught in the world of Christianity today. Paul, in his first book (letter) to Timothy said, "However the inspired utterance says definitely that in later periods of time (our time), some will fall away from the faith (truth), paying attention to "misleading" inspired utterances and teachings of demons, by hypocrisy of men who speak (teach) lies, marked in their conscience as with a branding iron." This shows clearly that demonic deception can "brand" an unsuspecting follower of false teachings for life. Once a person has been "branded" by demonic teachings, or Satan's "ministers," it is almost impossible to erase that "brand," even by the "true teachings from the scriptures. Most of the doctrines, customs and teachings in the "churches" today are misleading, deceptive lies. If what the "church leaders" are teaching is "truth," why so many different denominations? If all teach the truth, would all not be unified into "one denomination" not hundreds? In *Yahweh's* order of things there can be but one doctrine, one teaching, one truth and one Christian congregation. All others are demon inspired, and will be destroyed. Revelation 18:4 warns all who are branded by false religious teachings, "Get out of here my people, if you do not want to share with her (false religion) in her sins, and if you do not want to receive part of her plagues." An accurate knowledge of the truth is not something we should put trust in others for.

No, we should search for truth as for gold. Jesus counseled us, "Seek, and you will find, knock, and it will be opened to you, ask, and it will be given to you." This is the way to the "real truth."

If we are associated in anyway with false religion (Babylon the Great), we are again warned by an angel of *Yahweh*. He said, (Rev. 18:2); "And he (an angel) cried out with a strong voice, saying; she has fallen, Babylon the Great has fallen (been destroyed), and she has become a dwelling place of demons and a lurking place of every unclean and hated bird (Vultures). Does this religious harlot sound like an organization you would want to be part of? Has she "branded" you? Get out of her!

131

CHAPTER 15

DEMONIC POWERS AND INFLUENCE

Divination, from the Latin word, divus, means "pertaining to, or information received from or provided by demonic spirit beings." This includes and embraces a revelation of "secret knowledge" about future happenings or events through invoking spirit-like occult powers.

The various aspects included in the practice of divination are astrology, magic, sorcery, conjuration, fortune-telling, spiritism, witchcraft, voodooism, charming, hypnotism, the practice of secret rites, fetishism, and others. All who believe in, or practice divination are in reality, demonstrating belief in the superhuman powers of the demonic Gods. Practitioners of the various aspects of divination rely on demons to reveal the future to them through various means including; celestial phenomena, signs and omens, and "spirits of the dead", birds in flight, palmistry, divining rods, dreams, casting lots, crystal gazing, and many other forms of demonic communication.

All forms of divination, regardless of the name by which they may be called, is in sharp contrast, and is in direct defiance of *Yahweh's* word and purpose. He sternly and repeatedly warned, "Do not take up the practice of divination as the nations do, or anyone who employs magic or looks for omens, or anyone who binds others with a spell, or anyone who consults a spirit medium, or a professional foreteller of events of the future, or anyone who inquires or speaks with the dead. Anyone doing these things is found detestable to *Yahweh.*"

Satan and his demonic followers well know that mankind has a natural desire to foresee future events. Knowing this, he will use any device he can to gain man's trust and lead him away from his loyal worship of *Yahweh*. He also is the "only one" who knows the "end from the beginning," and must alienate man from God through demonic influence by making a "pretense of revealing" the future to man. Once man has been isolated from God because of his sinful spirit, he will then turn to demons as a substitute for his divine failures. *Yahweh* views the practice of any form of divination as comparable to the sin of rebellion against him, which incurs death.

Millions of people on earth today have turned to the secret arts and uncanny powers of demonism to gain the ability to accomplish supernatural occult acts. Through the use of spiritistic "black magic", some will employ special curses, spells, and the "evil eye", to harm or control one's enemies, while others turn to "white magic" in an effort to "break" such curses, or turning such curses back upon the originator. Also, through the use of magical formulas, practitioners of sorcery powers will attempt to influence people or alter future events.

Most of the concepts used in magic working sorcery has its basis in the belief that the demonic spirits can be persuaded to leave or enter a persons "psyche", or otherwise be controlled at the will of the magic practicing sorcerer.

Yahweh, in his written word, clearly condemned any and all forms of demon communication or worship. He said, "You must not practice magic, as for any man or woman in whom there proves to be a mediumistic spirit of prediction, they should be "put to death" without fail." *Yahweh* also warned all that he would cut short the days of all those who indulged in any form of occultism.

There have been numerous attempts by religious leaders to identify the "Anti-Christ," spoken of in God's word. They believe the Anti-Christ to be an individual person, which is expected to rise to political power in the "last days of this wicked system." However, the Bible record shows the Anti-Christ (man of lawlessness), to be not one man, but a composite organization made up of world supporters who stand in direct opposition to *Yahweh's* kingdom arrangement. Their denial of Jesus as the "Christ" and the Son of God, and their effort to replace Christ in "the sight of men" show that they are false prophets in disguise. The Anti-Christ (which occurs five times in the Bible record) includes all those who persecute the true followers of Christ. This composite "man of lawlessness" (or son of destruction), includes kingdoms, nations, and false religious organizations who revolt against *Yahweh* and his son, Christ Jesus. (Read 1st John, 4:2-3).

The expression, "man of lawlessness," as found in the Bible record is a pre-warning to true believers of a ruthless, hypocritical Anti-Christian apostasy that would engulf the entire religious system worldwide. Since this "man" is in fact a collective group of apostates, "He", is an earthly agent used by Satan and his "demon hordes" in their treasonous rebellion against *Yahweh's* right to universal sovereignty in his rule over the earth.

CHAPTER 16

THE DIVINE EXTRA-TERRESTRIAL

There's a master spirit being so technologically perfect that mankind cannot even comprehend the unbounded infinity of his existence. Man cannot even perceive the age of his being, because he has no beginning or starting point from which to measure the time of his beginning. He can therefore properly be called "the ancient of days," because his existence goes endlessly back into the very start of the ages in the past. He is also appropriately regarded as "the king of eternity," who counts millions of years as but "a watch in the night," and since this super extra-terrestrial is a "spirit being", his presence is not visible to the physical eyes of mankind. Any description of his appearance by man is mere conjectural speculation. He is so all-powerful even in appearance that he said; "No man can look upon my face and yet live."

Even so, this super-intelligent spirit being shows great love for the human family on

137

planet earth. He is the very personification of principled love. His unequaled love for man was clearly demonstrated by his willingness to sacrifice his only begotten son that we might regain eternal life. (His spirit son, born in flesh). *Yahweh's* love, especially for his "faithful" servants, is everlasting. It does not fail or diminish. His servant Paul exclaimed, "For I am convinced that neither death, nor life, nor "angels", nor governments, nor things now, nor things to come, nor powers, nor height, nor depth, nor any other creation can separate us from God's love that is in Christ Jesus our Lord."

Although, "no one" has seen *Yahweh's* face at anytime, he reveals himself as a "person", and figuratively speaks of his eyes, hands, heart, soul, feet, mind, and so on. He also speaks of showing feelings, having power, hearing, love, anger, and many other human-like attributes. His word and the visible things seen abundantly demonstrate the evidence that *Yahweh* is the creator of all things, and is indeed the highest God in the universe. The fantastic complexity of the reality of a superior technical intellect, "The very heavens are declaring the glory of God; And the work of his hands, the expanse is telling."

Yahweh's name is found more than 7000 times in the Hebrew scriptures alone. The use of his name so many times points to the fact that he is reality. He himself said, "I shall prove to be whatsoever I please." In other words, *Yahweh* is saying, "I will become whatever I wish to become in order to accomplish my will and purpose." He can also cause his spirit (or fleshly) creations to become whatever he wills, to carry out his righteous purposes. His will and purpose is so sure, that even the celestial bodies (stars), obey his commands, and cry out joyfully. He said, "So my word that goes forth from my mouth will prove to be. It will not return to me without results, but it will certainly do that which I have delighted, and it will have sure success in that for which I sent it." His word is so sure, with such certainty, that even great acts that appear to be impossible to humans are common in his viewpoint. Remember, his limitless power and intelligence makes all things possible to him, and no creation can turn his hand back.

Many people, who profess to believe in God, view him as an unexplainable "force," not as a spirit person. Others, who have sincerely examined the undeniable evidence accumulated through scientific investigation into the arrangement of the universe and the wonder of life on planet earth, can only conclude that such complex reality could only exist through the "will" of a "first cause." Even so, they still refuse to connect a superior authority to this "first cause." Design and creation requires extraordinary intelligence by this "first cause," while at the same time such intelligence requires a superior mind. Such a mind can only belong to person of "*Yahweh*" (Jehovah), which literally means, "He causes to become."

Since "*Yahweh*" is an uncreated spirit being (person), and has existed from "the beginning," he would clearly have a place of residence. His word informs us that "The heavens are his established place of dwelling." The Bible writer, Paul said, "Christ himself entered into heaven itself to appear before the person of God for us..." Although the heavens are identified as *Yahweh's* dwelling place, the heavens cannot contain the glory of his achievements. Even though he created hundreds of thousands of spirit beings (sons), not one is able to add anything to his superior knowledge or wisdom or contribute to his righteous standards. This of course is not to say *Yahweh* does not enjoy a pleasant relationship with his spirit sons. Having spirit bodies, angelic beings also have been given powerful capabilities and responsibilities in *Yahweh's* heavenly arrangement, and also make their abode in the heavens. They are called "Sons of the true God," or "Holy Ones."

The spirit realm is divinely organized into specific order and rank among the angelic creation. The foremost angel in both power and authority is identified as Michael (Jesus Christ) among the angelic host. He is the only archangel, and "all" other angelic beings are subject to his preeminence. The cherubs and seraphs are also among the highest ranking angels in privileges and honor. They have special duties and responsibilities among the "millions" of other angelic spirit beings. The "millions" of other angel's serve as messengers, guardians, and communicators between God and mankind. They also serve as executioners of *Yahweh's* will and purposes, whether it be protection, deliverance, or destruction.

Many uninformed may claim angels to be impersonal forces of energy sent to fulfill God's will. Not so! Individual angles are created as free agents, and like human creations, they can choose their own fate, and as humans, angels have distinct personalities, although on a much higher degree of intelligence and power.

These mighty angelic beings are not being sent to the earth to negotiate or confer with mankind regarding the coming social transformation of planet earth. "Michael," the archangel, now directs their visits to the earthly realm. God's favored servants on earth are assured that the largely invisible spirit beings protecting them are as "real" as the roman armies that surrounded Jerusalem before its destruction in the year 70 CE. The angels are also very active in the separation of the "wheat and the weeds" and the "sheep from the goats" taking place on the earth as you read this. Yes, the angel of "*Yahweh*" is camping out all around those who fear him, and he will stand side by side with Jesus Christ (Michael), as the final resolution is gained through the cleansing battle of Armageddon, after which, "The righteous themselves will possess the earth, and they will reside forever upon it (Ps. 37:29). Michael, the archangel, was given "all authority" over the entire creation including the millions of other spirit beings, the physical universe, and the earthly inhabitants since the year 1914 CE, when he was installed as king of kings and lord of lords by his father, "*Yahweh*". In the year 1919 CE Jesus Christ, assisted by his angelic host, instituted the separation activity spoken of by the Bible writer, Matthew, at Matthew 25:31-34, where Jesus said, "When the son of man (Jesus), arrives in his glory, (is installed as king), and all the "angels" with him, then (1919CE), he will sit down on his glorious throne (takes up his authority as king over the earth), all the nations (the kingdom of mankind), will be gathered before him (be given a final choice through a worldwide preaching work), and he (remember, he is the "word" of God), will "separate" people (his word does this separating), one from another, just as a shepherd separates the sheep (who accept his word), from the goats (who reject the good news), and he will put the sheep on his right hand (meaning a position of favor), but the goats on his left (a position of rejection), then the king will say to those on his right, "Come, you who have been blessed by "my father", inherit the kingdom (on earth) prepared for you from the founding of the world.

Here we clearly see how *Yahweh* gives everyone a choice of gaining his favor or disfavor. (life or death). If we wish to survive the coming transformation of the earth and live on into the one thousand-year reign of Christ, it is absolutely urgent that we seek his truth now. *Yahweh* is a merciful and loving extra-terrestrial spirit being. However, he has set an "appointed time" for his stated purpose, and his word is as definite as his eternal existence.

ABOUT THE AUTHOR

Mr. Charles P. Bost retired after a thirty-year career with the United States Air Force, Department of Defense in Civil/Mechanical Engineering. He was assigned to the Atomic Energy Commission, with the weapons development and testing division at Eglin Air Force Base, Fort Walton Beach, Florida, in 1949, 25th Drone Group (B-17 Aircraft).

In 1951-2 Mr. Bost participated in the first thermonuclear (hydrogen) bomb test at Enwitok Island in the Marshall Islands, as an instrumentation recovery and readout evaluation engineer. He continued in the weapons design and testing field, which led to the development of the "Smart" weapon series and participated in design research and flight test integration of "Standoff" weapons (Air research and development command).

Mr. Bost became aware of and interested in the UFO enigma in 1947 after the reported crash of an extra-terrestrial spacecraft on the Brazel property near Roswell, New Mexico. At the time of this incident, Mr. Bost was assigned to the strategic air command, 43rd Bomb Group at Davis Monthan Air Force

Base, Tucson, Arizona, piloting B-29 and B-50 aircraft. He has pursued a study of this phenomenon since that time. He has investigated scores of individual sightings for more than thirty years, and is fully convinced that UFO phenomena is real and has a direct connection with the angelic spiritual realm.

Mr. Bost is an aircraft pilot with more than 8000 flying hours logged, and is an eyewitness of "unknown" flying craft operating in earth's atmosphere and outer-space. He is a current member of the Mutual UFO Network (MUFON).

His studies and research of the UFO reality points to the revelation that planet earth is soon to be transformed and made into a "new creation."

www.ingramcontent.com/pod-product-compliance
Lightning Source LLC
Chambersburg PA
CBHW032018170526
45157CB00002B/759